THE COMPLETE PRE 11+ MATHS COURSE BOOK (YEAR 4)

Suitable for UK Grammar and Independent Schools

Advance Educational Services LTD

authorHOUSE

AuthorHouse™ UK
1663 Liberty Drive
Bloomington, IN 47403 USA
www.authorhouse.co.uk
Phone: UK TFN: 0800 0148641 (Toll Free inside the UK)
        UK Local: (02) 0369 56322 (+44 20 3695 6322 from outside the UK)

Published by AuthorHouse 01/18/2022

ISBN: 978-1-6655-9570-4 (sc)

Contents

Section 3

Answers   www.advanceeducation.co.uk/shop

Section 1
Topic 1 : Place Values

OBJECTIVES

In this module, you will be able to:

- ❖ Read whole numbers including large numbers
- ❖ Write whole numbers in words and figures
- ❖ Order and compare numbers

KEYWORDS:

Digit: These are the figures 0,1,2,3,4,5,6,7,8 or 9 used in numbers. E.g in 3152, there are 4 digits.

Place Value: The value of a digit relates to its position or place in a number. E.g. in 1481, the digits represent thousands, hundreds, tens and units respectively.

Previous knowledge:

- ➢ n/a

PLACE VALUE

We use only the digits 0, 1, 2, 3, 4, 5, 6, 7, 8 and 9 to write all our numbers. The position or **place** of a digit in a number is very important. This is because the place tells us the **value** of the digit.

e.g. The digit 3 in 62**3** is worth a different value to the 3 in **1**35

The place values are:

Ten Thousands	Thousands	Hundreds	Tens	Units
10Th	Th	H	T	U

Example 1

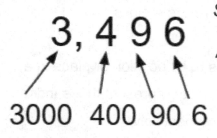

*Sometimes you'll see a comma in some numbers! This is used so we can read the number easier. A comma will be used **every three** numbers from the right to left.*

Examples: 78,087 and 5,742

The place values of 3496 are:

10Th	TH	H	T	U
0	3	4	9	6

So, we say 3496 is:

Three thousand (**TH**), four hundred (**H**) and ninety (**T**) six (**U**)

Example 2

What is the value of each digit in 86,015?

10Th	TH	H	T	U
8	6	0	1	5

Using the table, we can see that 86,015 is made up of:

8 lots of 10 thousands   = 80,000
6 thousands   =   6,000
0 hundreds   =   0
1 ten   =   10
5 units   =   5

So we say that 86,015 is:

Eighty six thousand and fifteen

You Try:

1.   Express the following words as numbers:

a)   One thousand three hundred and sixty two
b)   Sixty thousand, four hundred and eleven
c)   Five hundred and six

2.   Look at the number 65,452.

a)   What is the digit **4** worth?
b)   What is the digit **6** worth?
c)   What is the digit **2** worth?

3 Write down the value of the digits in **bold**, in words and in figures.

Each number has two of its digits **emboldened**.

Example: **32,507**	**Answer**	3: thirty thousand (30,000) 2: two thousand (2000) 5: five hundred (500)

a	5320	b	18,491	c	56,122
d	45,400	e	76,543	f	35,100

4 Copy out the numbers and for each number, underline the digit with the value in brackets.

Example 10,822 (eight hundred)	**Answer** 10,8̲22

a	88,282 (eighty thousand)	b	20,231 (two hundred)
c	30,400 (four hundred)	d	33,235 (thirty)
e	25,000 (five thousand)	f	83,838 (eight hundred)

5 In each shape, write the numbers in order, from the smallest to the largest.

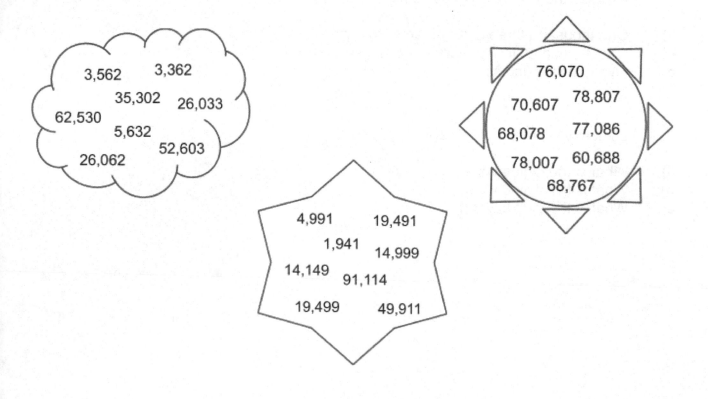

6 Write these numbers as figures:

a	thirty thousand	b	six thousand and ninety six
c	eight hundred	d	two hundred and two
e	twenty thousand	f	Thirty thousand and thirty three

7 Write these numbers in words:

a) 2,824

b) 4,806

c) 692

d) 29

e) 5,295

f) 71,491

g) 20,195

h) 51,239

i) 9,004

j) 40,252

k) 95,321

l) 4,920

m) 88,194

n) 2,495

o) 1,023

p) 492

q) 914

HINT : Remember to use the Place Value Table to help you!

EXAMPLE : Write 8,345 in words

1. Fill out the Place Value Table from right to left.

10Th	TH	H	T	U
0	8	3	4	5

2. Write the value of each number using the letters to help.

Eight **Thousand**, Three Hundred, Four **Tens** and Five **Units**

3. Combine the values above to form your answer.

Eight Thousand, Three Hundred and Forty Five

8. Write these numbers in figures:

a) Six hundred and thirty nine

b) Nine hundred and four

c) Seven hundred and seventy three

d) Two hundred and nineteen

e) Forty thousand and nine

f) Fifty eight thousand two hundred and one

g) Nine thousand and sixty two

h) Three hundred and three

i) Seventy thousand and thirty seven

j) Two hundred and sixty three

k) Ninety one

l) Three hundred and sixty four

m) Eight thousand two hundred and nine

n) Forty thousand one hundred and sixty four

o) Seven hundred and four

p) Ten thousand one hundred and seven

q) Forty three

HINT : Remember to use the Place Value Table to help you!

EXAMPLE: Write three hundred and ninety two in figures

1. Fill out the Place Value Table by breaking down the words.

Three **hundred** means that a '3' will go under the 'H' in our table.

Use this technique for the rest!

10Th	TH	H	T	U
0	0	3	9	2

2. Write your final answer down. In this case, we have 392!

9. Complete the table and write each distance in figures, the first one has been done for you.

Place	Distance to London in **words** (kilometres)	Distance to London in **figures** (kilometres)
1. Cardiff	Two hundred and sixty four	*264*
2. Bristol	One hundred and eighty three	
3. Manchester	Two hundred and eighty six	
4. Penzance	Five hundred and twenty one	
5. Edinburgh	Six hundred and forty two	
6. Nottingham	Two hundred and fourteen	
7. Land's End	Five hundred and five	
8. Leeds	Three hundred and ten	
9. Lincoln	Two hundred and eleven	
10. Aberdeen	Eight hundred and forty	

10. Complete the table and write each Footballer's value in words, the first one has been done for you.

Name	Value in **figures**	Value in **words**
1. Sam	£59,201	*Fifty nine thousand, two hundred and one pounds*
2. Alex	£91,405	
3. John	£24,392	
4. Pond	£10,015	
5. Terry	£29,502	
6. Smith	£50,194	
7. Elliott	£68,281	
8. Parks	£95,102	
9. Larry	£69,123	
10. Lucas	£80,486	

Section 1
Topic 2 : Number Patterns

OBJECTIVES

In this module, you will be able to:

❖   Remember the names of some common number patterns.

❖   Identify common number patterns (odd, even, prime and square numbers).

KEYWORDS:

Sequence   A sequence is a set of numbers arranged in order according to a **rule**. Each number in a sequence is called a **term**.
e.g. **2, 5, 8, 11**   the rule is 'add 3', the first 4 terms are 2, 5, 8 and 11.

Previous knowledge:

➢   n/a

EVEN

- A number that can be divided exactly by 2.
- Any number ending in 0, 2, 4, 6 or 8.

ODD

- A number that cannot be divided exactly by 2.
- Any number that ends in 1, 3, 5, 7 or 9.

You Try:

1. Write each of the following numbers and write down whether they are odd or even:

a) 415   f) 632,261

b) 1,890   g) 356,325

c) 12,489   h) 5,513

d) 89,897   i) 406,111

e) 759,588   j) 912,572

k) Copy these numbers and circle which ones are even.

28   46   83   1   5   64   30   86   77

l) Write all the odd numbers between 80 and 100.

m) Write all the even numbers between 50 and 80.

FACTORS

● A factor is a number that divides into another.

For example, the factors of 4 are numbers that divide into 4.

These are 1, 2 and 4.

● Factors come in pairs.

Example: List the factors of 12.

Answer: 1 x 12

2 x 6

3 x 4

The factors of 12 are: 1, 2, 3, 4, 6, and 12.

You Try:

2. Find the factors of the following numbers;

a)	6	i)	32
b)	8	j)	30
c)	16	k)	45
d)	24	l)	54
e)	36	m)	60
f)	28	n)	19
g)	42	o)	22
h)	18	p)	35

MULTIPLES

- Multiples are numbers that appear in the times table of a number. e.g. 2, 4, 6, 8 are multiples of 2 because they are in the 2 times table.

<u>**You Try**</u>:

3. **a)** Write down the first five multiples of 3

    b) 24 is a multiple of which numbers?

    c) Write the multiples of 4 between 20 and 53.

    d) What is the 10ᵗʰ multiple of 6?

    e) List the first ten multiples of 12.

    f) What is the next even multiple of 7 after 28?

    g) List all the even multiples of 9 between 1 and 100.

    h) List all the odd multiples of 11 between 1 and 132.

i) Which of the following are multiples of 6?

| 17 | 18 | 26 | 32 | 42 | 12 |

j) Which of the following are multiples of 7?

| 21 | 18 | 17 | 77 | 24 | 35 |

k) Which of the following are multiples of 8?

| 81 | 71 | 86 | 32 | 56 | 18 |

l) Which of the following are multiples of 9?

| 72 | 3 | 54 | 56 | 76 | 81 |

PRIME NUMBER

- A number that has only two numbers that divide into it (itself and 1). i.e. prime numbers **only** have **2 factors**.

  e.g. 2 , 3 , 5 are all prime.

  5 is **prime** because its only factors are 5 (itself) and 1.

- 1 is **not** prime.

You Try:

4. Draw and complete the following table for **1 to 20** and identify the prime numbers.

Number	Factors	Prime – 'Yes' or 'No'
1		
2		
3		
4		
5		
6		
7		
8		
9		
10		
11		
12		
13		
14		

SQUARE NUMBERS

- These numbers are found by multiplying a number by itself.

  e.g. 36 is a square number – it is the square of 6. In other words, 6 squared is the same as 6x6 which equals 36.

- The square of a number is written with a small 2 after the number as follows: 3^2 means '3 squared' = 3x3 = 9. 6^2 means '6 squared' = 6x6 = 36. The numbers 9 and 36 are called square numbers.

- The small '2' is called the **power**.

  It tells us how many times we should multiply the number by itself.

- **Square numbers** are numbers whose dots can be arranged as a square. 9 is a square number because:

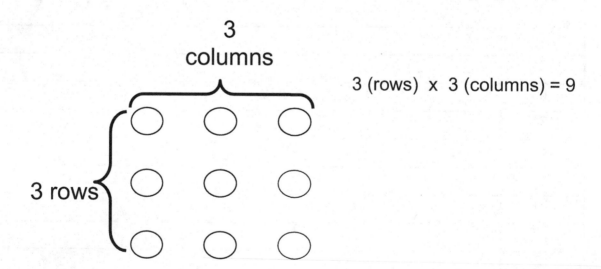

3 (rows) x 3 (columns) = 9

The first 12 square numbers are:

Square Numbers		Square Numbers	
1^2	1	7^2	49
2^2	4	8^2	64
3^2	9	9^2	81
4^2	16	10^2	100
5^2	25	11^2	121
6^2	36	12^2	144

SQUARE ROOT

● This is the opposite of a square number. When you are asked for the square root of a number, you have to find the number which when multiplied by itself gives you that original number. The symbol is $\sqrt{}$.

e.g. The square root of 36 is 6 and is written as $\sqrt{36} = 6$.
This is because 6 x 6 = 36.

You Try:

5. Write the following numbers then find its square root:

Example: 121 $\sqrt{121} = 11$

a) 4   f) 49   k) 100

b) 9   g) 64   l) 169

c) 16   h) 144   m) 81

d) 25   i) 196   n) 225

e) 36   j) 121

SEQUENCES

6.   Write the rule for the following sequences and write the next three terms:

<u>Example</u>:   7, 10, 13, 16.
$$+3 \quad +3 \quad +3$$

The rule is +3 and the next three terms are: 19, 22, 25

a)   3, 5, 7, 9

b)   5, 10, 20, 40

c)   53, 50, 47, 44

d)   8, 13, 18, 23

e)   256, 128, 64, 32

f)   3, 6, 12, 24

g)   52, 50, 48, 46

h)   1, 11, 21, 31

i)   1, 4, 9, 16

j)   1, 2, 4, 7

k)   2, 5, 10, 17

l)   3, 6, 11, 18

m)   2, 8, 18, 32

n)   2, 5, 9, 14, 20

HINT:
The rule may not be the same each time! Think about the pattern between each step to help you.

Answer these exam questions

a) Is the number 126,854 odd or even?

b) I want to find the answers to adding any two numbers. They are either odd or even. Find the answers:

i) even + even

ii) even + odd

iii) odd + even

iv) odd + odd

c) Are these numbers multiples of 2? Write in full sentences.

| 582 | 673 | 105 | 942 | 86 | 88 |

What does this tell you about the type of number it is?

d) There is only one prime number that is <u>even</u>.

What is this number?

Why can't the other even numbers be prime numbers?

e) Copy this sentence. 20, 650, 100 and 240 all end in zero. They all have _____ as a factor.

This means that numbers that end in _____ cannot be prime.

f) Copy this sentence. 25, 35, 75, 105 and 215 all end in five. They all have _____ as a factor.

This means that numbers that end in _____ cannot be prime.

g) Work out these squares.

Draw dots arranged as a square for them.

i) 8 x 8

ii) 12 x 12

iii) 7 x 7

h) Which of the following numbers are factors of 18?

4   9   6   10   12   18   2   1   5   7

i) Which of the following numbers are factors of 56?

8   2   9   5   4   10   12   13   18

j) What are the next three terms in the sequence: 32, 16, 8, 4?

k) List the factors of 48.

l) Write the 5th multiple of 4.

m) Write all the square numbers below 40.

n) What is the square of 7?

o) Write the rule for the sequence: 49, 45, 41, 37.

p) What is the square root of 81?

q) List all the prime numbers between 20 and 40.

r) A sequence has the rule, 'double the number and take away 1'. The first number is 2. Write the next 4 terms.

s) Find:

i) $\sqrt{64}$ ii) $\sqrt{121}$

Section 1
Topic 3: Rounding

OBJECTIVES

In this module, you will be able to:

❖ Round numbers to the nearest 10, 100 and 1000.

❖ Estimate sums by rounding.

KEYWORDS

Rounding: This is a way of making numbers simpler.

e.g. using rounding, change 19 + 12 to 20 + 10.

Estimate: This means to have a guess using rounded numbers.

e.g. the estimate of 19 + 12 is around 30.

Approximate: The same as estimating.

Previous knowledge:

➢ Place Values

ROUNDING TO THE NEAREST 10

To round a number to the nearest 10, look at the **UNITS** digit.

If the units digit is

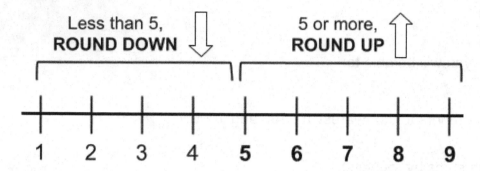

Less than 5, **ROUND DOWN**

5 or more, **ROUND UP**

1 2 3 4 5 6 7 8 9

Examples:

1)   Round 356 to the nearest 10.

The last digit in 356 is 6. We round it <u>up</u> to <u>360</u>.

2)   Round 352 to the nearest 10.

The last digit of 352 is 2. We round it <u>down</u> to <u>350</u>.

3)   Round 475 to the nearest 10.

475 ends in a 5.

Since we always round a 5 <u>up</u>, 475 to the nearest 10 is <u>480</u>.

46 → 50
6 is **more** than 5! So 46 will **ROUND UP**

22 → 20
2 is **less** than 5! So 46 will **ROUND DOWN**

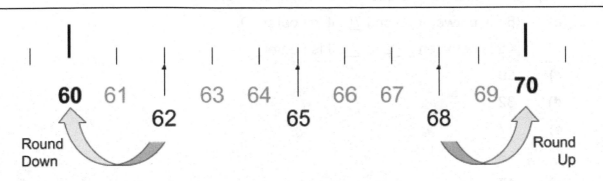

62 is between 60 and 70 but it is closer to 60 than 70.
We round 62 **down**.
You say 62 rounded to the nearest 10, is <u>60</u>.

68 is also between 60 and 70 but it is closer to 70
We round 68 **up**.
You say 68 rounded to the nearest 10, is <u>70</u>.

<u>You Try:</u>

1. Round these numbers to the nearest 10.

a)	43		**m)**	659
b)	27		**n)**	285
c)	24		**o)**	164
d)	35		**p)**	187
e)	65		**q)**	295
f)	98		**r)**	102
g)	3		**s)**	574
h)	122		**t)**	3,345
i)	399		**u)**	5,688
j)	925		**v)**	1,236
k)	391		**w)**	4,857
l)	742		**x)**	3,989

2. Round off these numbers to the nearest ten.

Copy and complete the questions like the example.

a) 64 is between <u>60</u> and <u>70</u>. It is nearer <u>60</u>.

b) 45 is between __ and __. It is nearer ___.

c) 56

d) 32

e) 17

f) 58

g) 47

h) 38

i) 29

j) 42

k) 82

l) 86

m) 57

n) 78

o) 27

p) 85

q) 38

r) 66

s) 99

t) 26

u) 32

v) 84

w) 56

x) 60

y) 96

z) 8

3. Round these numbers to the nearest 10:

a)	60	**i)**	83
b)	96	**j)**	63
c)	8	**k)**	47
d)	36	**l)**	26
e)	98	**m)**	52
f)	45	**n)**	73
g)	68	**o)**	24
h)	6	**p)**	6

ROUNDING TO THE NEAREST 100

To round to the **nearest 100,** look at the **TENS COLUMN.** If the TENS digit is:

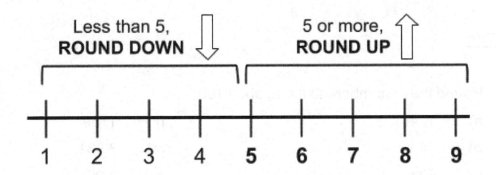

Examples:

1. Round 3281 to the nearest 100.

The tens digit in 3281 is 8. We round it <u>up</u> to 3300.

2. Round 3216 to the nearest 100.

The tens digit of 3216 is 1. We round it <u>down</u> to 3200.

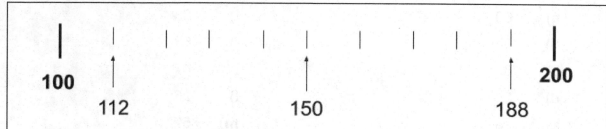

100   200

112   150   188

112 is between 100 and 200 but it is closer to 100 than 200, so **round down**. You say 112 rounded to the nearest 100, is <u>100</u>.

188 is also between 100 and 200 but it is closer to 200, so **round up**. You say 188 rounded to the nearest 100, is <u>200</u>.

150 is halfway between 100 and 200. **Round up**. You say 150 rounded to the nearest 100, is <u>200</u>.

You Try:

4. Round these numbers to the nearest 100:

a)	124	**l)**	1759
b)	740	**m)**	3,831
c)	698	**n)**	2,958
d)	148	**o)**	3,702
e)	47	**p)**	2,058
f)	270	**q)**	6,807
g)	450	**r)**	2,945
h)	2,098	**s)**	9,819
i)	4,286	**t)**	8,421
j)	8,050	**u)**	1,856
k)	3,829	**v)**	28,12

5.   Copy and fill in the spaces like the example:

   a)   285 is between <u>200</u> and <u>300</u>. It is closer to <u>300</u>.

   b)   592 is between ___ and ___, It is closer to ___.

   c)   682

   d)   8,711

   e)   129

   f)   815

   g)   474

   h)   8,237

   i)   912

   j)   1,019

   k)   178

   l)   826

   m)   5,542

   n)   806

   o)   7,470

   p)   4,587

   q)   1,248

   r)   16,402

   s)   18,231

   t)   49,212

   u)   93,148

   v)   74,21

ROUNDING TO THE NEAREST 1000

Look at the rules for rounding to the nearest 10 and 100. Can you work out the rule for rounding to the nearest 1000?

To round a number to the nearest 1000, look at the **HUNDREDS COLUMN**. If the HUNDREDS <u>digit</u> is:

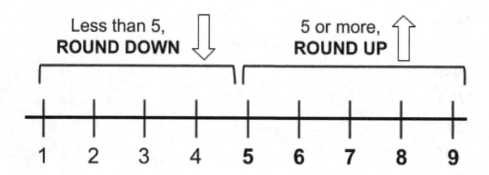

<u>**Examples**</u>:

1. Round 4,559 to the nearest 1000.

    The hundreds digit in 4,559 is 5. We round it <u>up</u> to <u>5,000</u>.

2. Round 4,295 to the nearest 1000.

    The hundreds digit of 4,295 is 2. We round it <u>down</u> to <u>4,000</u>.

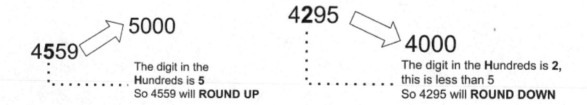

You Try:

6.   Use the rule to round these numbers to the nearest 1000:

a) 998

b) 1,050

c) 448

d) 2,315

e) 1,999

f) 1,299

g) 2,567

h) 23,123

i) 18,543

j) 13,489

k) 19,195

l) 85,129

m) 17,532

n) 52,305

o) 82,449

p) 73,793

q) 89,912

r) 58,291

s) 17,582

t) 48,823

u) 49,012

v) 8,582

w) 4,567

x) 8,280

y) 9,485

z) 5,500

7.   Round the numbers as indicated in the brackets.

- 32 (10)
- 10,139 (100)
- 1,568 (100)
- 755 (100)
- 3185 (10)
- 6729 (1000)
- 568 (1000)
- 20,405 (1000)

- 15,662 (10)
- 448 (10)
- 948 (10)
- 15,420 (1000)
- 725 (10)
- 77,637 (100)
- 12,225 (100)

8.   Round each of the number to nearest: i) 10   ii) 100   iii) 1000

a)   8,234

b)   23,139

c)   12,225

d)   25,042

e)   7,417

f)   1,781

g) 980

h) 20,906

i) 18,562

j) 9,310

k) 3,205

l) 7,76

9. Copy and complete the following table in your book.

	Nearest 10	Nearest 100
628		
809		
848		
992		
408		
204		
190		

	Nearest 10	Nearest 100	Nearest 1000
1,202			
1,025			
1,053			
21,555			
25,555			

ESTIMATING

When you estimate the answer to a problem, you:

- First round the numbers.

- Then find the answer.

<u>**Example 1**</u>:   Estimate the answer to 21 + 39.

<u>**Answer**</u>:   We round 21 to 20, and 39 to 40.

Then we have: 20 + 40 = <u>60</u>.

Rounds down to 20 ⬇ $\frac{21}{20}$ $\frac{40}{39}$ ⬆ **Rounds down to** 40   **20 + 40 = 60**

<u>**Example 2**</u>:   My Mum saw pairs of socks for £2.99 each.

She bought 4 pairs of socks.

How much did she spend?

<u>**Answer**</u>:   Round £2.99 to £3.

She buys <u>4</u> pairs of socks:

£3 x 4 = <u>£12</u>

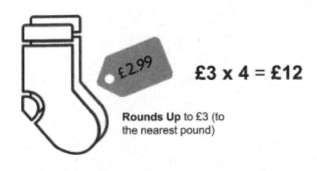

Rounds Up to £3 (to the nearest pound)

You Try:

10. Use rounding to estimate the following sums:

a) 22 + 9 + 42

b) 74 – 48

c) 14 x 9

d) 79 ÷ 9

e) 119 + 289

f) 84 – 19 – 12

g) 24 + 59 + 11

h) 93 – 47 – 43

i) 38 + 5 + 17

j) 63 – 54 - 5

k) 91 – 18 – 49

l) 14 + 48 + 35

m) 38 x 99

n) 184 ÷ 18

o) 73 x 7

p) 58 ÷ 29

q) 6 x 75

r) 99 ÷ 47

11. An office chair costs £27.99. How much do:

a) 2 chairs cost

b) 5 chairs cost?

12. In a toy store, Pandas were the most popular in the afternoon.

Janis sold 17 of them.

Each Panda cost £1.45.

Use rounding to estimate how much money she received.

13. A box of chocolate on Mother's Day costs £4.99.

Approximately how many boxes can I buy for £30?

14. A pipe has length 28m. I divide it into 5 pieces. Roughly how long is each piece?

Section 1
Topic 4: Adding and Subtracting Whole Numbers

OBJECTIVES

In this module, you will be able to:

- ❖ Add numbers using traditional method with 'carrying'.
- ❖ Subtract numbers using traditional method with 'borrowing'.

KEYWORDS

Partition: The splitting up of numbers into tens, units etc.

Example: **19** = $\underline{10}$ + $\underline{9}$ **72** = 70 + 2.

Therefore: **19 + 72** = $\underline{10}$ + 70 + $\underline{9}$ + 2

$= 80$ $+ 11$

$= \underline{91}$

Decomposition: The borrowing from one place value to a place value below it.

Example:

```
  5   0              45  10
-  3  4    becomes  -  3   4
```

Previous knowledge:

- ➢ Place Values

In worded problems, the question may use **these words** instead of "add" or "subtract". Here are some examples:

WORDS FOR ADDING	Example
Add	4 <u>add</u> 5 is 9.
Altogether	<u>Altogether</u>, 14 and 5 make 19.
Increase	<u>Increase</u> 14 by 5 makes 19.
More	19 is 5 <u>more</u> than 14.
Plus	5 <u>plus</u> 14 is 19.
Sum	The <u>sum</u> of 5 and 14 is 19.
Total	The <u>total</u> of 5 and 14 is 19.

WORDS FOR SUBTRACTING	Example
Minus	17 <u>minus</u> 3 is 14.
Subtract	<u>Subtract</u> 3 from 17 gives 14.
Difference	The <u>difference</u> between 17 and 3 is 14.
Take away	17 <u>take away</u> 3 is 14.
Fewer	3 <u>fewer</u> than 17 is 14.
Less	Abi has 17 apples and 14 oranges. How many <u>less</u> oranges does she have?
How many less/ How much more	Rebecca has £14. <u>How much more</u> is needed to get £17?
Reduce	<u>Reduce</u> 17 by 3.

Addition and **subtraction** are the **_opposite_** of each other.

Example 1: $\underline{12} + 3 = 15$

$\qquad 15 - 3 = \underline{12}$

Notice how I have 12 again after adding 3 and then subtracting 3.

Example 2: $\underline{19} - 4 = 15$

$\qquad 15 + 4 = \underline{19}$

Again, I now have 19 again after subtracting 4 and then adding 4.

Strategies for mental addition and subtraction

General

Count forward or back in repeated steps of 10, 5 or 2.

Adding two similar numbers in value

Split the larger one up into the smaller number plus the difference.

e.g. $12 + \underline{13} = 12 + \underline{12 + 1} = 25$

Separate into tens and units

e.g. $\underline{28} + \mathbf{36} = \underline{20} + \underline{8} + \mathbf{30} + \mathbf{6}$

$\qquad = \underline{20} + \mathbf{30} + \underline{8} + \mathbf{6}$

$\qquad = 64$

Add/subtract 9, 19, 29 (or 11, 21, 31)

Add or subtract 10, 20, 30 and then adjust by 1.

e.g. $\quad \underline{49} + \mathbf{31} = \underline{50} - \underline{1} + \mathbf{30} + \mathbf{1}$

$\qquad = \underline{50} + \mathbf{30} - \underline{1} + \mathbf{1}$

$\qquad = 80$

You Try:

1. Find the answers to the following.

   Remember to show your working and to write in your books.

   a) What is the sum of 4784 and 628?

   b) Total 16362 and 9020.

   c) Add together 10184, 492 and 8920.

   d) What number is 511 more than 9168?

   e) What is the difference between 6581 and 31094?

   f) Subtract 4065 from 12208.

   g) What is 7886 less than 9822?

   h) Take 8118 away from 14238.

   i) There are 866 girls and 925 boys in gym club.

   How many children are in the club altogether?

   j) Imran has 5903 football stickers.

   He loses 389.

   How many stickers has he got left?

   k) The attendances at three football matches were 45674, 49128 and 38763.

   What is the total attendance for the three games?

   l) There were 23491 people at the concert in all the Stands.

   In Stand A, there were 13432 people.

   In Stand B, there were 7652 people.

   How many people were in Stand C?

ADDING WITH CARRYING

Question: Add 578 to 27.

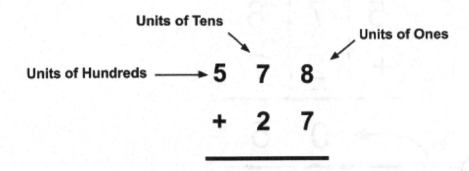

Answer:

Step 1:   Start by adding together the Units: 8 + 7 = **15**.

Step 2:   Write the 5 underneath in the Units column.

'Carry' the **1** from the **15**, and write it underneath the Tens:

$$5 \quad 7 \quad \vdots 8 \vdots$$
$$+ \quad 2 \quad \vdots 7 \vdots$$
$$5$$
$$1$$

Step 3:   Add the Tens: 7 + 2 = 9.

Don't forget to add on the carried **1** underneath: 9 + 1 = <u>10</u>.

Step 4:   Write the 0 underneath the Tens.

                'Carry' the <u>1</u> and write it underneath in the Hundreds column.

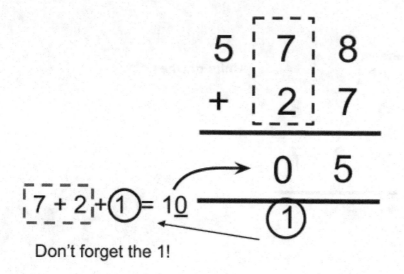

$7 + 2 + 1 = 1\underline{0}$

Don't forget the 1!

Step 5:   Add the Hundreds including the <u>1</u> underneath: 5 + 1 = 6

                Write the 6 in the Hundreds column.

```
      5 : 7 : 8
    + :  2 : 7
    _____
      6   0   5
      1       1
```

5 + ①= 6

Don't forget
the 1!

You Try:

2.   Write the sums in your book. Use the adding with carrying method.

a)
```
    3 4
+   1 9
_____
```

b)
```
    5 2
+   8 5
_____
```

c)
```
    3 7
+   4 8
_____
```

d)
```
    8 6
+   5 4
_____
```

e)
```
    8 3
+   9 6
_____
```

f)
```
    6 3
+   3 5
_____
```

g)
```
  9 4 0
+   6 4
_____
```

h)
```
  4 8 4
+   2 7
_____
```

i)
```
  6 4 8
+   2 2
_____
```

j)
```
  6 6 3
+   4 1
_____
```

k)
```
  8 5 7
+   4 8
_____
```

l)
```
  3 4 1
+   6 7
_____
```

m)
```
  3 2 5
+ 1 9 7
_____
```

n)
```
  4 2 4
+ 3 1 3
_____
```

o)
```
  2 5 6
+ 9 7 8
_____
```

p)
```
  3 2 1 5
+   7 9 4
_____
```

q)
```
  1 3 8 4
+   4 0 7
_____
```

r)
```
  2 0 4 8
+   5 5 6
_____
```

s)
```
  5 3 7 5
+ 2 0 3 4
_____
```

t)
```
  3 0 0 2
+ 1 9 9 9
_____
```

u)
```
  2 1 6 3
+ 2 4 1 8
_____
```

SUBTRACTION WITH BORROWING

Question: 136 – 59

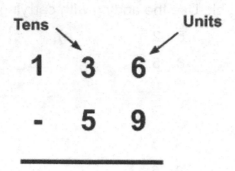

Step 1:   Look at the Units: we cannot do 6 – 9.

We need **to borrow one** from the 3 in the Tens column.

Cross out the 3 because it will become 2 (3 - 2 = 1).

Add the borrowed 1 above the 6 in the Units to give 16.

The new subtraction is now 16 – 9 = 7.

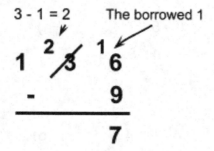

Step 2:   Look at the Tens: we cannot do 2 – 5.

We need to borrow from the 1 in the Hundreds column.

Cross out the 1 because it will become 0.

Add the borrowed 1 to the 2 in the Hundreds to give 12.

The new subtraction is now 12 – 5 = 7

You Try:

3. Write the subtractions in your book.

   Use the **subtracting with borrowing** method to work out the answers.

a)
```
    7 4
  -   6 4
  --------
```

b)
```
    8 4
  -   7 8
  --------
```

c)
```
    4 0
  -   2 3
  --------
```

d)
```
    6 5
  -   4 7
  --------
```

e)
```
    4 8
  -   2 9
  --------
```

f)
```
    4 1
  -   3 7
  --------
```

g)
```
    7 8 5
  - 4 2 7
  --------
```

h)
```
    9 2 0
  - 4 5 3
  --------
```

i)
```
    7 0 0
  - 6 0 2
  --------
```

j)
```
    8 5 0
  - 6 4 0
  --------
```

k)
```
    8 9 1
  - 7 9 9
  --------
```

l)
```
    8 0 0
  - 6 2 9
  --------
```

m)
```
    6 4 8
  - 4 7 7
  --------
```

n)
```
    9 4 7
  - 3 9 9
  --------
```

o)
```
    8 0 0 0
  - 7 8 9 1
  ----------
```

p)
```
    5 5 2 2
  - 1 8 7 7
  ----------
```

q)
```
    6 0 2 0
  - 2 3 4 7
  ----------
```

r)
```
    4 0 0 0
  - 2 9 1 3
  ----------
```

s)
```
    6 2 9 7
  - 1 4 6 8
  ----------
```

t)
```
    7 1 2 1
  - 2 1 4 6
  ----------
```

u)
```
    3 0 0 1
  - 1 7 8 0
  ----------
```

Section 1
Topic 5: Multiplying – Simple and Long

OBJECTIVES

In this module, you will be able to:

❖   Multiply two and more digit numbers by a single digit (short multiplication).

❖   Multiply two and more digit numbers by 2-digit numbers (long multiplication).

KEYWORDS:

WORDS FOR MULTIPLYING	Example
Multiply	Multiply 7 with 3 is 21.
Times	3 times 7 is 21.
Product	The product of 7 and 3 is 21.
Double	Double 3 is 6.
Groups of	3 groups of 7 is 21.
Lots of	3 lots of 7 are 21.
Sets of	3 sets of 7 is 21.

Previous knowledge:

➢   Place Values
➢   Addition and Subtraction

SHORT MULTIPLICATION

Here are two methods you can use to multiply. Use the one that you feel comfortable with and can use accurately and quickly.

Example: 126 x 3

Method 1:   Traditional method

● Multiply 3 with 6 = 18.

Carry the 1 and write it underneath in the Tens column.

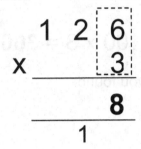

```
  1 2 6
x     3
_____
      8
  1
```

● Multiply 3 with 2 = 6.

Add the carried 1 from before: 6 + 1 = 7

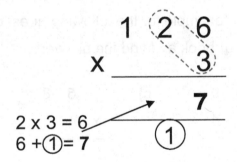

```
  1 2 6
x     3
_____
      7
```

2 x 3 = 6
6 + ①= 7   ①

● Multiply 3 with 1 = 3.

```
  1 2 6
x     3
_____
  3 7 8
    1
```

Method 2:   Grid method

- Break each number into Hundreds, Tens and Units.
  Here, 126 is split into 100, 20 and 6.

- Draw the grid and multiply each of the numbers:

x	100	20	6
3	**300**	**60**	**18**

$$\text{100 x 3} = \mathbf{300}$$

- Now add up all the numbers you found:

$$300 + 60 + 18 = \underline{378}$$

You Try:

1. Practice the methods in the example for the following questions.
   Copy the questions into your book and find the answers.

a)
```
    6 8
x     5
_____
_____
```

b)
```
    4 9
x     4
_____
_____
```

c)
```
    5 8
x     6
_____
_____
```

d)
```
    6 8
x     7
_____
_____
```

e)
```
  7 0 4
x       3
_____
_____
```

f)
```
  4 8 2
x       4
_____
_____
```

g)
```
  3 6 0
x       8
_____
_____
```

h)
```
  1 2 7
x       6
_____
_____
```

i)
```
  1 0 7
x       6
_____
_____
```

j)
```
  3 5 9
x       5
_____
_____
```

k)
```
  2 4 7
x       8
_____
_____
```

l)
```
  2 3 6
x       2
_____
_____
```

m)
```
    8 4 2
x       3
_____

_____
```

n)
```
    7 6 3
x       9
_____

_____
```

o)
```
    5 5 5
x       7
_____

_____
```

p)
```
    4 5 6
x       3
_____

_____
```

q)
```
  2 3 4 1
x       6
_____

_____
```

r)
```
  3 0 5 6
x       8
_____

_____
```

s)
```
  2 8 5 9
x       9
_____

_____
```

t)
```
  2 6 1 0
x       5
_____

_____
```

LONG MULTIPLICATION

Here are two methods you can use to multiply. Use the one that you feel comfortable with and can use accurately and quickly.

Example: 126 x 34

Method 1:   Traditional method

Step 1:

● Multiply 4 by 6 = 24.

Carry the 2 and write it underneath in the Tens column.

```
    1 2 6
x     3 4
_____
        4
_____
      2
```

- Multiply 4 by 2 = 8.

  Add the carried 2 from before: 8 + 2 = 10.

  Now carry the 1 and write it underneath in the Hundreds column.

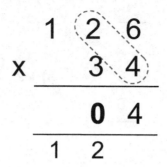

```
    1  2  6
 x     3  4
 _____
       0  4
    1  2
```

- Multiply 4 by 1 = 4.

  Add the carried 1 from before: 4 + 1 = 5.

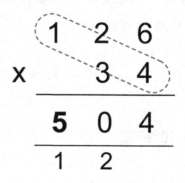

```
    1  2  6
 x     3  4
 _____
    5  0  4
    1  2
```

Step 2:

- Remove the carried numbers from before and the question looks something like this:

```
    1  2  6
 x     3  4
 _____
    5  0  4
```

- Write a 0 in the next row in the Units column as follows.

  This is because we are now looking at the <u>3 tens</u> of 34.

```
    1  2  6
x      3  4
-----------
    5  0  4
          0
```

- Multiply 3 with 6 = 18.

  Write the carried 1 underneath in the Hundreds column.

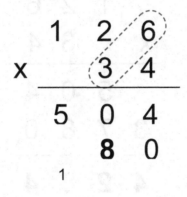

```
    1  2  6
x      3  4
-----------
    5  0  4
       8  0
    1
```

- Multiply 3 by 2 = 6.

  Add the carried 1 from before: 6 + 1 = 7

```
    1  2  6
x      3  4
-----------
    5  0  4
    7  8  0
    1
```

- Multiply 3 by 1 = 3.

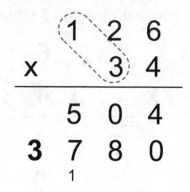

```
      1  2  6
  x      3  4
  _____
      5  0  4
  3   7  8  0
         1
```

- Now add our two lines together: 504 + 3780 = <u>4284</u>

```
      1  2  6
  x      3  4
  _____
      5  0  4
  3   7  8  0
  _____
  4   2  8  4
```

Method 2: Grid method.

- Break each number into Hundreds, Tens and Units.

  126 is split into 100, 20 and 6.

  34 is split into 30 and 4.

- Draw the grid and multiply each of the numbers:

x	100	20	6
30	**3000**	**600**	**180**
4	**400**	**80**	**24**

- Now add up all the numbers you found:

  3000 + 400 + 600 + 80 + 180 + 24 = <u>4284</u>

You Try:

2. Practice the methods in the example for the following questions.

  Copy the questions into your book and find the answers.

a)   4 3
  x 3 2
  ‾‾‾‾‾‾

b)   6 8
  x 2 1
  ‾‾‾‾‾‾

c)   3 8
  x 5 4
  ‾‾‾‾‾‾

d)   6 2
  x 4 5
  ‾‾‾‾‾‾

e)   2 4
  x 3 5
  ‾‾‾‾‾‾

f)   4 3
  x 8 1
  ‾‾‾‾‾‾

g)   3 4
  x 7 9
  ‾‾‾‾‾‾

h)   2 7
  x 4 6
  ‾‾‾‾‾‾

i)   3 1
  x 4 9
  ‾‾‾‾‾‾

j)   7 2
  x 3 9
  ‾‾‾‾‾‾

k)   4 5
  x 7 2
  ‾‾‾‾‾‾

l)   1 9
  x 8 5
  ‾‾‾‾‾‾

m)	3 8 8	**n)**	2 0 9	**o)**	8 7 1	**p)**	5 7 4
	x 2 3		x 1 4		x 7 5		x 2 6

q)	1 0 7	**r)**	9 8 3	**s)**	3 4 5	**t)**	1 7 4
	x 6 8		x 4 1		x 5 2		x 3 8

3.   Solve these problems and show your working.

a)   Milk crate contains 24 bottles of milk.
There are 32 crates on a milk float.
How many bottles are there on the milk float?

b)   A school organises a trip for Year 4.
The tickets cost £9.
How much will it cost for 34 children?

c)   My 5 friends have 85 marbles each.
How many have they got in total?

d)   Arfan drives 18 miles every day to get to and from work.
He works 5 days a week.
How many miles does he drive for his job?

Year 4 Pre 11+ Maths

Autumn Half Term Test

Full Name: …………………… Test Date: ………………………………

Tutor: ………………………… Day of Lessons: ……………………

This test lasts 30 minutes.

Read the questions carefully and try all the questions.
If you cannot do a question, move on to the next one.
If you have finished, check over your answers.

The total marks for this paper is 40.

1.   Look at the number 67,295.

a)   Write this number in words.

_____ (1)

b)   What is the value of the 7?

_____ (1)

c)   What is the digit of the ten thousands value?

_____ (1)

2.   Underline the digit with the value in the bracket for the following:

a)   22,192 (twenty thousand)   (1)

b)   50,454 (four hundred)   (1)

c)   97,119 (one ten)   (1)

3.

a)   What is three thousand and ninety one in figures?

_____ (1)

b)   What is ninety thousand and four in figures?

_____ (1)

4. Use these numbers to answer the questions.

6   45   27   31   16

a) Write all the odd numbers.   _____ (1)

b) Which of these is a factor of 36?   _____ (1)

c) Which of these is a multiple of 5?   _____ (1)

d) Which of these is a square number?   _____ (1)

e) Which of these is a prime number?   _____ (1)

5.

a) What is 8^2?   _____ (1)

b) Work out $\sqrt{196}$.   _____ (1)

c) Fill in the missing numbers for the sequence.   (3)

7, 14, 21, ____, 35, ____, ____, 56, 63

1, 2, 4, 7, ____, ____, 22, 29, 37, 46, ____

6. For her homework, Rebecca wrote:

"The first five prime numbers are 1, 2, 3, 5 and 7".

Is she correct? Explain your answer.   (2)

_____

_____

_____

7. For his homework, Harry could not do the following problem:

> Use rounding to estimate this calculation:
>
> 28 + 743 + 2595

His teacher suggested doing the following questions to help:

a) Round 28 to the nearest 10: _____ (1)

b) Round 743 to the nearest 100: _____ (1)

c) Round 2595 to the nearest 1000: _____ (1)

d) Use your answers to <u>estimate</u> the calculation.

_____ (1)

8. Here is another question from Harry's homework.

The answer is wrong and he could not work out why.

$$
\begin{array}{r}
2\ 5\ 4 \\
\times\quad\ \ 3 \\
\hline
6\ 6\ 2 \\
\hline
{\scriptstyle 1}
\end{array}
$$

a) Which of the digits in the answer is wrong?

_____ (1)

b) What is the correct digit from part a)?

_____ (1)

9.

a) Find the difference between 7298 and 4909. (3)

b) Find the total of 5621 and 2806. (3)

c) A packet of sweets has 25 sweets in them.
For the party, we bought 37 packets.
How many sweets will we have for the party? (3)

10. At the grocery store, they sold oranges for 28p each.

a) How much would it cost to buy 7 oranges? (2)

b) My friend Janice decided to buy 18 oranges.
How much did she pay for her oranges? (3)

Section 1
Topic 7 : Dividing

OBJECTIVES

In this module, you will be able to:

❖ Divide two or more digit numbers by a single digit (short division).

❖ Divide two or more digit numbers by 2-digit numbers (long division)

KEYWORDS:

<u>WORDS FOR Dividing</u>	<u>Example</u>
Divide	<u>Divide</u> 21 by 3 equals 7.
Share	<u>Share</u> 21 between 3 gives 7.
Split	<u>Split</u> 21 evenly between 3 gives 7.
Group	<u>Group</u> 21 into 3 gives 7 in each group.
Half	<u>Half</u> 24 means 24 divided by 2, which gives 12.
Remainder	This is the amount left after dividing. e.g. 15 divided by 2 is 7 remainder 1.

Previous knowledge:

➢ Place Values

BASICS

Dividing and multiplying are the <u>opposite</u>.

If you know your times tables, you may be able to work backwards:

From your times table: 8 x 7 = 56

Going backwards: 56 ÷ 7 = 8   and   56 ÷ 8 = 7

Division can be expressed in different ways:

$$96 \div 4 \qquad \frac{1}{4} \text{ of } 96 \qquad 4\overline{\smash{\big)}96} \qquad \frac{96}{4}$$

All the above mean "how many 4's make 96".

Example 1: 53 ÷ 7

*Imagine you were passing out **53 sweets** to your 7 friends. If you handed them out **one by one in a circle**, they would each have **7 sweets** and you would have **4 left**!*

Answer:

Write the multiples of 7 out: 7, 14, 21, 28, 35, 42, **49**, 56

The closest number below 53 is **49**.

49 and 4 makes 53.

This means 53 ÷ 7 = <u>7 remainder 4</u>.

Example 2: Find $\frac{1}{8}$ of 80.

How do we visualise $\frac{1}{8}$?

Think about splitting this bar into **8 equal pieces**.

80 is made up of:	80
2 parts of 40	40 40
or 4 parts of 20	20 20 20 20
or 8 parts of 10	10 10 10 10 10 10 10 10

Let's half it

Since 8 lots of '**10**' make 80, we say that

10 makes 1 out of 8 parts of 80. This is

written as $\frac{1}{8}$ of 80.

Answer:

$\frac{1}{8}$ of 80 is the same as 80 ÷ 8.

$\frac{1}{8}$ of 80 = 80 ÷ 8 = <u>10</u>

You Try:

1. Write your working out in your book and show working out.

a)   42 ÷ 7

b)   27 ÷ 3

c)   54 ÷ 9

d)   55 ÷ 5

e)   56 ÷ 8

f)   28 ÷ 4

g)   54 ÷ 6

h)   32 ÷ 8

i)   63 ÷ 9

j)   25 ÷ 5

2.

a) $\frac{1}{3}$ of 12

b) $\frac{1}{8}$ of 128

c) $\frac{1}{4}$ of 80

d) $\frac{1}{9}$ of 126

e) $\frac{1}{5}$ of 80

f) $\frac{1}{7}$ of 91

g) $\frac{1}{8}$ of 96

h) $\frac{1}{6}$ of 78

i) $\frac{1}{5}$ of 120

j) $\frac{1}{4}$ of 72

SHORT DIVISION

Traditional Method:

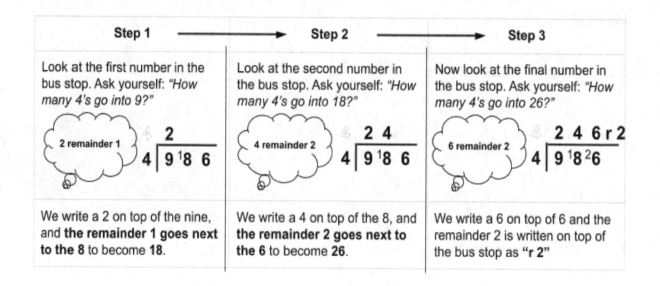

Step 1 ⟶	Step 2 ⟶	Step 3
Look at the first number in the bus stop. Ask yourself: *"How many 4's go into 9?"*	Look at the second number in the bus stop. Ask yourself: *"How many 4's go into 18?"*	Now look at the final number in the bus stop. Ask yourself: *"How many 4's go into 26?"*
We write a 2 on top of the nine, and **the remainder 1 goes next to the 8** to become **18**.	We write a 4 on top of the 8, and **the remainder 2 goes next to the 6** to become **26**.	We write a 6 on top of 6 and the remainder 2 is written on top of the bus stop as **"r 2"**

You Try:

3.

a) 2 | 6 5 4 b) 7 | 4 9 0 c) 6 | 1 5 9 0

d) 4 | 1 9 6 e) 3 | 5 8 8 f) 8 | 7 5 0 5

g) 7 | 3 7 4 h) 5 | 6 4 5 i) 7 | 9 9 9 4

j) 2 | 5 8 7 k) 2 | 2 5 4 l) 9 | 4 8 9 0

m) 3 | 7 5 0 n) 7 | 3 5 7 o) 2 | 4 5 9 1

4.

a. 7 sour sweets weigh 357g.
How much does 1 sweet weigh?

b. 220 cards were put into packs of 5.
How many complete packs were there altogether?

c. There are 240 children in Ludlow Junior School and 8 classes. How many children in each class?

d. The school kitchen buys 114 fish fingers and each child gets 3 for their lunch. How many children can have fish fingers?

e. The school librarian has £300 to spend on books.
Each book cost £6.
How many books can she buy?

f. Mrs Rose has 85 colouring pencils and wants to divide them equally between her 5 tables.
How many pencils will each table get?

g. Vi has 363 posters and they want to spread them out evenly across 3 different nearby towns. How many posters will go to each town?

h. Caitlyn is setting up cupcake stands for charity. She bakes 832 cupcakes and sells them evenly across 4 stands. How many cupcakes did each stand sell?

LONG DIVISION

<u>**Question**</u>: $435 \div 15$

<u>**Answer**</u>:

<u>Step 1</u>: $4 \div 15 = \underline{0}$ remainder 4 (ignore remainder).

          q) Write the <u>0</u> on top.

          r) Multiplying 15 with <u>0</u> gives 0.

          s) Write this 0 underneath the 4.

          t) Now subtract $4 - 0 = 4$ and write this below.

          u) Bring down the 3.

```
                0
    1  5  | 4   3   5
           ‾‾‾‾‾‾‾‾‾‾
        -    0   ↓
           ‾‾‾‾‾
             4   3
```

<u>Step 2</u>: $43 \div 15 = \underline{2}$ remainder 13 (ignore remainder).

          w) Write the <u>2</u> on top.

          x) Multiplying 15 with 2 gives 30.

          y) Write this 30 underneath the 43 from before.

          z) Now subtract $43 - 30 = 13$ and write this below.

          aa) Bring down the 5.

```
           4   0   2
    1 5  | 4   3   5
        -   0   |
           ‾‾‾‾ ↓
             4   3
         -   3   0  |
            ‾‾‾‾‾‾  ↓
             1   3   5
```

Step 3:   135 ÷ 15 = <u>9</u>

- Write the <u>9</u> on top.
- Multiplying 15 with <u>9</u> gives 135.
- Write this 135 underneath the 135 from before.
- Now subtract 135 – 135 = 0 and write this below.

```
        0 2 9
  15  | 4 3 5
   -    0 |
        ↓
        4 3
   -    3 0 |
           ↓
        1 3 5
   -    1 3 5
        ‾‾‾‾‾
            0
```

Therefore: 435 ÷ 15 = <u>29</u>

You Try:

5.   Copy and complete in your books.

a) 1 2 | 6 1 2   **b)** 1 3 | 6 7 6   **c)** 2 4 | 3 0 0 0

d) 1 6 | 7 2 0   **e)** 1 9 | 4 7 5   **f)** 2 7 | 1 1 0 7

g) 1 1 | 6 4 9   **h)** 1 4 | 6 5 8   **i)** 1 6 | 1 2 1 6

Section 1
Topic 8: Equivalent Fractions

OBJECTIVES

In this module, you will be able to:

❖   Write a <u>fraction</u> in both **figures and words**

❖   Label a given diagram with the correct fraction

❖   Shade correctly a diagram to represent a fraction

❖   Identify **equivalent fractions** and shade in diagrams

KEYWORDS:

Whole number	A number which does not contain any fractions e.g. 5, 44, 192.
Fraction	A number less than one whole number. e.g. $\frac{1}{2}, \frac{5}{8}, \frac{1}{3}$
Numerator	This is the top number of the fraction.
Denominator	This is the bottom number of the fraction.
One half	$\frac{1}{2}$
One third	$\frac{1}{3}$
One quarter	$\frac{1}{4}$
Equivalent Fractions	Different fractions that represent the same decimal or whole number. e.g. $\frac{1}{2} = \frac{2}{4} = 0.5$

Previous knowledge:

➤   Multiplication and Division

RECOGNISING FRACTIONS

A fraction consists of a top number and a bottom number.

The top number is called the numerator.

The bottom number is called the denominator.

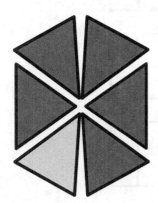

The line in between these two numbers represents a divide symbol:

$\frac{1}{2}$ means the numerator 1 is divided by the denominator 2.

Given a diagram that has been split into equal parts, we can label it with a fraction.

Here is a hexagon split into equal parts.

There are 5 dark regions and 1 light region. We can represent this in terms of fractions:

Dark regions: $\frac{5}{6}$ Five sixths

Light regions: $\frac{1}{6}$ One sixth

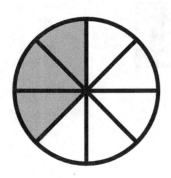

Here is a circle split into equal parts.

There are 3 shaded regions and 5 unshaded regions:

Shaded regions: $\frac{3}{8}$ Three eighths

Unshaded regions: $\frac{5}{8}$ Five eighths

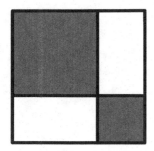

This diagram is a square not split into equal parts. Therefore we cannot represent this as easily with a fraction.

YOU TRY:

Write both in <u>figures</u> and <u>words</u> the fraction of each diagram that is shaded.

1.

a)

b)

c)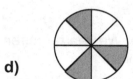

d)

e)

f)

g)

h)

i)

2.

a)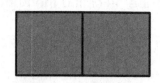

b)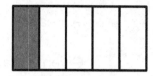

c)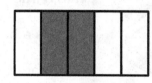

d)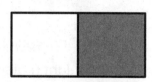

e)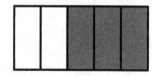

f)

g)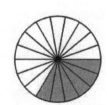

h)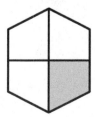

EQUIVALENT FRACTIONS

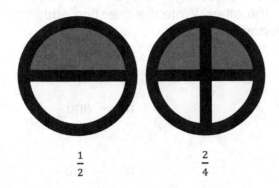

$$\frac{1}{2} \qquad \frac{2}{4}$$

➤ The two circles above have the <u>same amount</u> shaded.

➤ The fractions below correspond to the amount shaded.

➤ These two fractions are <u>equivalent fractions</u>.

➤ To convert from one another, we multiply both the top and bottom by the **same** number.

$$\overset{\times 2}{\overbrace{\quad}} $$
$$\frac{1}{2} = \frac{2}{4}$$
$$\underset{\times 2}{\underbrace{\quad}}$$

➤ To go the other way, we <u>divide</u> both the top and bottom by the **same** number.

$$\overset{\div \; 2}{\overbrace{\quad}}$$
$$\frac{1}{2} = \frac{2}{4}$$
$$\underset{\div \; 2}{\underbrace{\quad}}$$

➤ Some more examples of equivalent fractions are:

$$\frac{2}{5} = \frac{4}{10} \qquad \frac{1}{3} = \frac{4}{12} \qquad \frac{7}{9} = \frac{21}{27} \qquad \frac{4}{6} = \frac{15}{18}$$

NOW TRY:

Identify whether or not the following are equivalent fractions or not and write how to get from one to the other. Write the question and answer in your books. The first one has been done for you.

3.

a) $\frac{3}{4}$ and $\frac{6}{8}$ **Yes, x2**

b) $\frac{4}{5}$ and $\frac{7}{8}$

c) $\frac{1}{12}$ and $\frac{2}{24}$

d) $\frac{5}{9}$ and $\frac{7}{8}$

e) $\frac{5}{6}$ and $\frac{7}{10}$

f) $\frac{10}{20}$ and $\frac{1}{2}$

g) $\frac{4}{6}$ and $\frac{1}{3}$

h) $\frac{6}{11}$ and $\frac{3}{22}$

i) $\frac{9}{10}$ and $\frac{18}{20}$

j) $\frac{2}{6}$ and $\frac{1}{3}$

k) $\frac{4}{20}$ and $\frac{2}{10}$

l) $\frac{16}{32}$ and $\frac{1}{2}$

m) $\frac{5}{2}$ and $\frac{7}{4}$

n) $\frac{10}{20}$ and $\frac{1}{3}$

o) $\frac{4}{5}$ and $\frac{8}{10}$

p) $\frac{11}{44}$ and $\frac{1}{4}$

q) $\frac{6}{30}$ and $\frac{1}{6}$

r) $\frac{4}{15}$ and $\frac{1}{12}$

4.   **Copy** these diagrams and write your answers into your books.

In the first diagram, shade correctly the fraction indicated.

Then, shade in the correct amount in the second diagram to represent an equivalent fraction.

Write this equivalent fraction down.

The first one has been done for you.

a)

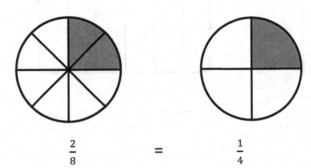

$$\frac{2}{8} \quad = \quad \frac{1}{4}$$

b)

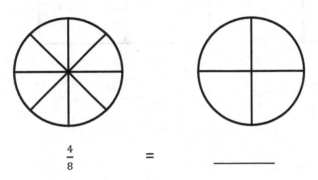

$$\frac{4}{8} \quad = \quad \underline{\hspace{2cm}}$$

c)

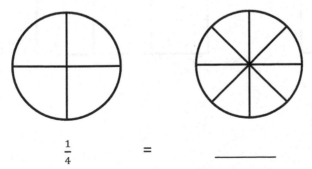

$$\frac{1}{4} \quad = \quad \underline{\hspace{2cm}}$$

d)

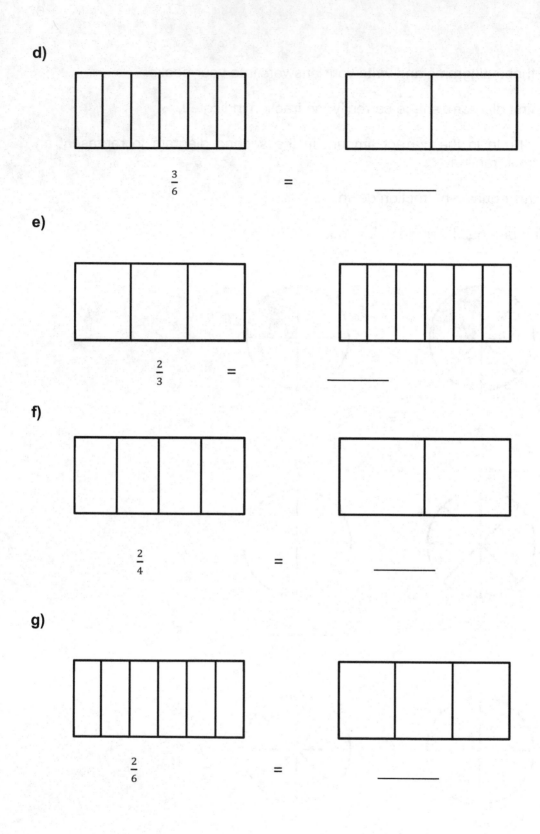

$$\frac{3}{6} \qquad = \qquad \underline{\hspace{2cm}}$$

e)

$$\frac{2}{3} \qquad = \qquad \underline{\hspace{2cm}}$$

f)

$$\frac{2}{4} \qquad = \qquad \underline{\hspace{2cm}}$$

g)

$$\frac{2}{6} \qquad = \qquad \underline{\hspace{2cm}}$$

Section 1
Topic 9 : Highest Common Factor (HCF) and Simplifying Fractions

OBJECTIVES

In this module, you will:

❖ Learn how to split a number into its factors

❖ Identify the highest common factor of two numbers

❖ Use the highest common factor to cancel fractions

KEYWORDS:

Factor   A number that divides another number.

Highest Common Factor   The highest number that divides both
                                    numbers.

Equivalent Fractions   Different fractions that represent the
                                    same decimal or whole number.

$$\text{e.g. } \frac{1}{2} = \frac{2}{4} = 0.5$$

Previous Knowledge:

❖ Multiplication and Division
❖ Equivalent Fractions

Factors

A number is a factor of another if it divides into it without a remainder.

Using division, we can easily find the factors of a number.
They come in <u>pairs</u>.

We can also use our times tables to help us.

<u>Example 1</u>:   Find the factors of 15.

<u>Answer</u>:
$$15 = 15 \times 1$$
$$15 = 5 \times 3$$

The factors of 15 are <u>1, 3, 5 and 15</u>.

The numbers 1, 3, 5 and 15 all divide 15.

<u>Example 2</u>:   Find the factors of 20.

<u>Answer</u>:
$$20 = 20 \times 1$$
$$20 = 10 \times 2$$
$$20 = 5 \times 4$$

The factors of 20 are <u>1, 2, 4, 5, 10 and 20</u>.

All these numbers divide into 20.

<u>Example 3</u>:   Find the factors of 24.

<u>Answer</u>:
$$24 = 24 \times 1$$
$$24 = 12 \times 2$$
$$24 = 8 \times 3$$
$$24 = 6 \times 4$$

The factors of 24 are <u>1, 2, 3, 4, 6, 8, 12 and 24</u>.

All these numbers divide 24.

Highest Common Factor (HCF)

We found the factors of the numbers 15, 20 and 24 before.

The <u>highest common factor</u> **(HCF)** is the <u>highest number</u> that appears in both sets of factors.

Example 1:   Find the highest common factor of 15 and 20.

Answer: The factors of: 15 are 1, 3, <u>5</u> and 15.

                                  20 are 1, 2, 4, <u>5</u>, 10 and 20.

             The highest number that appears in both is <u>5</u>.

             Therefore, **the highest common factor** is <u>5</u>.

Example 2:   Find the highest common factor of 20 and 24.

Answer: The factors of: 20 are 1, 2, <u>4</u>, 5, 10 and 20.

                                  24 are 1, 2, 3, <u>4</u>, 6, 8, 12 and 24.

             The highest number that appears in both is <u>4</u>.

             Therefore, **the highest common factor** is <u>4</u>.

YOU TRY:

1. For each of the following,

    i)   Find all <u>the factors</u> of two numbers.

    ii)   Find the <u>highest common factor</u> of the two numbers.

a)   9 and 18

b)   16 and 20

c)   10 and 15

d)   16 and 24

e)   5 and 20

f)   19 and 57

g)   12 and 9

h)   49 and 21

i)   20 and 10

j)   35 and 21

k)   16 and 30

l)   8 and 15

m)   6 and 33

n)   32 and 18

o)   12 and 15

p)   15 and 18

q)   42 and 16

r)   15 and 25

s)   23 and 46

Cancelling Fractions

'<u>Simplify</u>' means to reduce the top and bottom numbers in our fraction to their <u>lowest</u> values.

We can use the <u>highest common factor</u> to simplify.

<u>Example 1</u>:   Simplify $\frac{15}{20}$.

<u>Answer</u>:   The highest common factor of 15 and 20 is 5.

     Dividing both the numerator and denominator by 5.

$$\overset{\div 5}{\overbrace{\frac{15}{20}}} = \underset{\div 5}{\underbrace{\frac{3}{4}}}$$

Example 2:   Reduce the fraction $\frac{20}{24}$ to its lowest terms.

Answer: The highest common factor of 20 and 24 is 4.

Divide both the top and bottom by 4 to simplify.

$$\frac{20}{24} = \frac{5}{6}$$

YOU TRY:

2.   For the following, use your answers in Question 1.

Simplify the fraction in each question.

a) $\frac{9}{18}$

b) $\frac{16}{20}$

c) $\frac{10}{15}$

d) $\frac{16}{24}$

e) $\frac{5}{20}$

f) $\frac{19}{57}$

g) $\frac{9}{12}$

h) $\frac{21}{49}$

i) $\frac{10}{20}$

j) $\frac{21}{35}$

k) $\frac{16}{30}$

l) $\frac{8}{15}$

m) $\frac{6}{33}$

n) $\frac{18}{32}$

o) $\frac{12}{15}$

p) $\frac{15}{18}$

q) $\frac{16}{42}$

r) $\frac{15}{25}$

s) $\frac{23}{46}$

t) $\frac{14}{26}$

Section 1
Topic 10: Improper Fractions and Mixed Numbers

OBJECTIVES

In this module, you will be able to:

❖ Cancel down <u>improper fractions</u> into <u>whole numbers</u>

❖ Change <u>improper fractions</u> into <u>mixed numbers</u>

❖ Change <u>mixed numbers</u> into <u>improper fractions</u>

KEYWORDS:

Improper Fraction A fraction where the numerator is greater than the denominator e.g. $\frac{5}{2}, \frac{9}{7}, \frac{15}{2}$

Also called a top heavy fraction.

Mixed Number A number that has a whole number and a fraction e.g. $1\frac{2}{3}, 5\frac{4}{5}, 4\frac{9}{11}$

Previous knowledge:

❖ Multiplication and Division

Improper Fractions

Improper fractions (also known as top heavy fractions) are fractions whose numerator is larger than its denominator.

Examples: $\frac{5}{4}, \frac{9}{7}, \frac{15}{2}$

5 is larger than 4 so this is an improper fraction (or top heavy) fraction

$\frac{5}{4}$

5 → Numerator

4 → Denominator

Simplifying Improper Fractions into Whole Numbers

Recall that the line in fractions means division. Therefore, all we need to do is divide the numerator by the denominator:

Examples: $\frac{20}{4}$ means $20 \div 4 = 5$

$\frac{9}{3}$ means $9 \div 3 = 3$

We can visualise $\frac{20}{4}$ by using this diagram

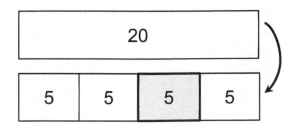

If we split a bar of 20 into 4 equal parts, each part will be 5. This is because 5 x 4 = 20.

YOU TRY:

1. Write these improper fractions as whole numbers.

a) $\dfrac{10}{2}$

b) $\dfrac{20}{5}$

c) $\dfrac{6}{2}$

d) $\dfrac{8}{4}$

e) $\dfrac{9}{3}$

f) $\dfrac{15}{5}$

g) $\dfrac{18}{9}$

h) $\dfrac{20}{2}$

i) $\dfrac{35}{7}$

j) $\dfrac{42}{6}$

k) $\dfrac{39}{3}$

l) $\dfrac{16}{4}$

m) $\dfrac{52}{13}$

n) $\dfrac{4}{1}$

o) $\dfrac{19}{19}$

p) $\dfrac{33}{11}$

q) $\dfrac{60}{12}$

r) $\dfrac{90}{30}$

s) $\dfrac{45}{9}$

t) $\dfrac{100}{50}$

Changing Improper Fractions into Mixed Numbers

A <u>mixed number</u> is a number with both a whole number and a fraction together.

To write an <u>improper fraction</u> as a <u>mixed number</u>, we divide similarly to before.

The remainder will make it into a mixed number. We write **the remainder as the numerator** and write the same denominator as in the question.

<u>Example 1</u>:   Write $\frac{16}{5}$ as a mixed number.

<u>Answer</u>: $\frac{16}{5}$ means 16 ÷ 5 = <u>3</u> with <u>remainder 1</u>.

This means we have 3 wholes and 1 remainder that couldn't be divided, this remains as a fraction.

So $\frac{16}{5}$ can be written as a mixed number:

$$\frac{16}{5} = 3\frac{1}{5}$$

<u>Example 2</u>:   Write $\frac{9}{4}$ as a mixed number.

<u>Answer</u>: $\frac{9}{4}$ means 9 ÷ 4 = <u>2</u> with <u>remainder 1</u>.

This means we have 2 wholes and 1 remainder that couldn't be divided, this remains as a fraction.

So $\frac{9}{4}$ can be written as a mixed number:

$$\frac{9}{4} = 2\frac{1}{4}$$

YOU TRY:

2. Simplify these improper fractions into mixed numbers.

a) $\dfrac{14}{3}$

b) $\dfrac{19}{2}$

c) $\dfrac{7}{3}$

d) $\dfrac{5}{3}$

e) $\dfrac{20}{6}$

f) $\dfrac{8}{5}$

g) $\dfrac{8}{3}$

h) $\dfrac{17}{2}$

i) $\dfrac{22}{4}$

j) $\dfrac{17}{5}$

k) $\dfrac{20}{3}$

l) $\dfrac{23}{2}$

m) $\dfrac{23}{7}$

n) $\dfrac{16}{3}$

o) $\dfrac{20}{19}$

p) $\dfrac{16}{5}$

q) $\dfrac{27}{12}$

r) $\dfrac{37}{30}$

s) $\dfrac{28}{9}$

t) $\dfrac{57}{50}$

Changing Mixed Numbers into Improper Fractions

Example 1:   Change $2\frac{7}{9}$ into an improper fraction.

Answer: First we multiply the whole number, 2, by the denominator, 9:

$$2 \times 9 = \underline{18}$$

Then add this to the numerator of the fraction:

$$\underline{18} + 7 = 25$$

This is the new numerator.

Keeping the denominator the same, we get $\frac{25}{9}$.

Example 2:   Change $3\frac{1}{4}$ into an improper fraction.

Answer: First we multiply the whole number 3 by the denominator 4:

$$3 \times 4 = \underline{12}$$

Then add this to the numerator of the fraction:

$$\underline{12} + 1 = 13$$

This is the new numerator.

Keeping the denominator the same, we get $\frac{13}{4}$.

YOU TRY:

3. Change these mixed numbers into improper fractions.

a) $2\frac{8}{11}$

k) $3\frac{2}{9}$

b) $3\frac{5}{6}$

l) $1\frac{1}{30}$

c) $2\frac{2}{3}$

m) $3\frac{1}{2}$

d) $4\frac{4}{5}$

n) $4\frac{1}{9}$

e) $5\frac{1}{2}$

o) $2\frac{1}{3}$

f) $1\frac{3}{5}$

p) $3\frac{3}{5}$

g) $2\frac{1}{2}$

q) $6\frac{2}{5}$

h) $3\frac{1}{3}$

r) $1\frac{5}{12}$

i) $2\frac{2}{7}$

s) $1\frac{1}{11}$

j) $2\frac{2}{5}$

Section 1
Topic 11: Lowest Common Multiple (LCM)

OBJECTIVES

In this module, you will be able to:

❖ Find <u>multiples</u> of a number

❖ Find the **<u>lowest common multiple</u>** of two numbers

❖ Use lowest common multiple to write fractions with the same denominators

❖ Order fractions in terms of size

KEYWORDS:

Multiple	Multiples of a number are its times table.
Lowest Common Multiple	This is the smallest number that appears as a multiple of two numbers.
>	The number on the left is <u>greater than</u> the number on the right
<	The number on the left is smaller <u>than</u> the number on the right

---

Previous knowledge:

❖ Multiplication and Division
❖ Equivalent Fractions

Multiples

Multiples of a number are its times tables.

Examples: The multiples of 2 are 2, 4, 6, 8, 10 and so on.

                   The multiples of 3 are 3, 6, 9, 12, 15 and so on.

Lowest Common Multiple (LCM)

The lowest common multiple of a pair of numbers is the smallest factor that appears **in both** of the sets of multiples.

Example 1:   Find the LCM of 4 and 5.

Answer: The multiples of 4 are 4, 8, 12, 16, **20**, 24, 28....

                   The multiples of 5 are 5, 10, 15, **20**, 25, 30....

                   The smallest factor for both 4 and 5 is **20**.

                   This means the LCM is **20**.

Example 2:   Find the LCM of 3 and 9.

                   The multiples of 3 are 3, 6, **9**, 12, 15, 18, 21....

                   The multiples of 9 are **9**, 18, 27, 36, 45, 54....

                   The smallest factor for both 3 and 9 is **9**.

                   This means the LCM is **9**.

YOU TRY:

1. Find the lowest common multiple of these two numbers:

a)	3 and 5	**g)**	8 and 9
b)	5 and 6	**h)**	11 and 6
c)	4 and 9	**i)**	4 and 6
d)	3 and 8	**j)**	3 and 27
e)	7 and 6	**k)**	9 and 12
f)	6 and 8	**l)**	7 and 56
		m)	12 and 11

Using the LCM to find Equivalent Fractions

We have already looked at equivalent fractions.

Using the <u>LCM</u>, we can write a pair of fractions with the <u>same denominator</u>.

<u>Example 1</u>: Write equivalent fractions for $\frac{2}{5}$ and $\frac{1}{2}$ so that they both have the same denominator.

<u>Answer</u>:

➤ Find the LCM of the denominators:

The multiples of 5 are 5, <u>10</u>, 15, 20, 25....

The multiples of 2 are 2, 4, 6, 8, <u>10</u>, 12....

The LCM of 2 and 5 is <u>10</u>.

➤ 10 is the <u>second</u> multiple of 5.

So we multiply the top and bottom of $\frac{2}{5}$ by 2:

x2

$$\frac{2}{5} = \frac{4}{10}$$

x2

➤ 10 is the <u>fifth</u> multiple of 2.

So we multiply the top and bottom of $\frac{1}{2}$ by 5:

x5

$$\frac{1}{2} = \frac{5}{10}$$

x5

➤ Now we have two <u>equivalent fractions</u>:

$$\frac{2}{5} = \frac{4}{10} \text{ and } \frac{1}{2} = \frac{5}{10}$$

Using > and < to compare Fractions

We use > to tell us the number on the left is <u>greater than</u> the number on the right.

We use < to tell us the number on the left is <u>smaller than</u> the number on the right.

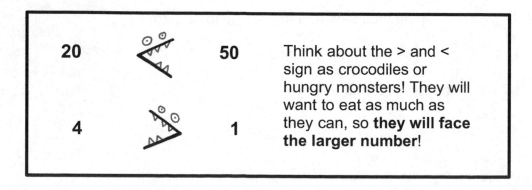

20	<	**50**
4	>	**1**

Think about the > and < sign as crocodiles or hungry monsters! They will want to eat as much as they can, so **they will face the larger number!**

Example: Which one of these are bigger: $\frac{2}{5}$ or $\frac{1}{2}$?

Answer: Using the LCM of 2 and 5 before, we found:

$$\frac{2}{5} = \frac{4}{10} \text{ and } \frac{1}{2} = \frac{5}{10}$$

It is clear to see that $\frac{5}{10} > \frac{4}{10}$

because 5 *is greater than 4*

so 5 > 4

Therefore: $\frac{1}{2} > \frac{2}{5}$

YOU TRY:

2.   For the following, use the method in Question 1. Copy each question and write > or < in the blank spaces.

a) $\frac{1}{2}$ $\frac{1}{4}$

b) $\frac{1}{3}$ $\frac{1}{6}$

c) $\frac{1}{7}$ $\frac{1}{14}$

d) $\frac{1}{2}$ $\frac{3}{4}$

e) $\frac{2}{3}$ $\frac{1}{6}$

f) $\frac{1}{3}$ $\frac{4}{6}$

g) $\frac{1}{5}$ $\frac{6}{10}$

h) $\frac{2}{5}$ $\frac{1}{10}$

i) $\frac{3}{5}$ $\frac{3}{10}$

j) $\frac{4}{5}$ $\frac{9}{10}$

k) $\frac{4}{5}$ $\frac{11}{15}$

l) $\frac{1}{3}$ $\frac{1}{5}$

m) $\frac{1}{2}$ $\frac{4}{7}$

n) $\frac{3}{5}$ $\frac{3}{4}$

o) $\frac{3}{4}$ $\frac{3}{7}$

p) $\frac{2}{3}$ $\frac{5}{6}$

q) $\frac{1}{5}$ $\frac{11}{20}$

r) $\frac{2}{5}$ $\frac{6}{10}$

s) $\frac{3}{5}$ $\frac{10}{15}$

t) $\frac{4}{5}$ $\frac{11}{12}$

u) $\frac{7}{9}$ $\frac{3}{4}$

v) $\frac{9}{10}$ $\frac{12}{15}$

Using LCM to order fractions

To order fractions easily, we can use the LCM to get the <u>same denominator</u> for <u>all fractions</u>.

<u>Example:</u>   Order these fractions starting with the smallest.

$$\frac{1}{2} \qquad \frac{2}{3} \qquad \frac{1}{4} \qquad \frac{7}{12}$$

<u>Answer:</u>

➢ Find the LCM of all the denominators:

Multiples: 2:   2, 4, 6, 8, 10, <u>12</u>...
                       3:     3, 6, 9, <u>12</u>, 15, 18...
                       4:     4, 8, <u>12</u>, 16, 20...
                       12:   <u>12, 24, 28,....</u>

The LCM of 2, 3, 4 and 12 is <u>12</u>.

➢ Rewrite every fraction so that the denominator is 12:

12 is the <u>sixth multiple</u> of 2 so multiply top and bottom of

$\frac{1}{2}$ by 6:

$$\frac{1 \times 6}{2 \times 6} = \frac{6}{12}$$

12 is the <u>forth multiple</u> of 3 so multiply the top and bottom of

$\frac{2}{3}$ by 4:

$$\frac{2 \times 4}{3 \times 4} = \frac{8}{12}$$

12 is the <u>third multiple</u> of 4 so multiply the top and bottom of

$\frac{1}{4}$ by 3:

$$\frac{1 \times 3}{4 \times 3} = \frac{3}{12}$$

12 is the <u>first multiple</u> of 12 so $\frac{7}{12}$ it stays the same.

> Now we have: $\dfrac{6}{12}, \dfrac{8}{12}, \dfrac{3}{12}, \dfrac{7}{12}$

> Ordering them by the numerator: $\dfrac{3}{12}, \dfrac{6}{12}, \dfrac{7}{12}, \dfrac{8}{12}$

> Now write the original fractions: $\dfrac{1}{4}, \dfrac{1}{2}, \dfrac{7}{12}, \dfrac{2}{3}$

3. Order these fractions starting with the smallest.

a) $\dfrac{1}{3}, \dfrac{2}{5}, \dfrac{7}{15}$

b) $\dfrac{1}{4}, \dfrac{15}{16}, \dfrac{7}{8}$

c) $\dfrac{2}{3}, \dfrac{8}{9}, \dfrac{14}{18}, \dfrac{5}{6}$

d) $\dfrac{3}{4}, \dfrac{5}{6}, \dfrac{2}{3}, \dfrac{5}{12}$

e) $\dfrac{5}{7}, \dfrac{19}{28}, \dfrac{11}{14}, \dfrac{50}{56}$

4. Order these fractions starting with the largest fraction.

a) $\dfrac{5}{9}, \dfrac{16}{18}, \dfrac{29}{36}, \dfrac{2}{3}$

b) $\dfrac{4}{7}, \dfrac{4}{5}, \dfrac{30}{35}, \dfrac{10}{35}$

c) $\dfrac{1}{3}, \dfrac{4}{5}, \dfrac{3}{4}, \dfrac{7}{10}$

d) $\dfrac{5}{7}, \dfrac{3}{4}, \dfrac{11}{28}, \dfrac{9}{14}$

Year 4

Autumn End of Term Test

Full Name: …………………. Test Date: ……………….…............

Tutor: ……………….………. Day of Lessons: ……………..………

This test lasts 30 minutes.

Read the questions carefully and try all the questions.
If you cannot do a question, move on to the next one.
If you have finished, check over your answers.

The total marks for this paper is 40.

Do not write in this space

1.

a) 504 sweets are shared between 8 children.
How many sweets does each child get? (2)

b) What is eighty thousand **minus** two thousand nine hundred and twenty one? (3)

c) Find the **sum** of 10731, 89 and 6107 (3)

d) Find the **product** of 39 and 274. (3)

2. Have a look at the number:

41,608

a) Write this number in words. (1)

_____

b) Write the value of the digit 6 in this number. (1)

_____

c) Round 41,608 to the nearest **100**. (1)

_____

d) Round 41,608 to the nearest **1000**. (1)

_____

3. At random, Karen picked these numbers out of the hat:

64   37   48   12

She was then asked the following questions.

a) Which of these numbers is a **multiple** of 24? _____ (1)

b) Which of these numbers is a **factor** of 24? _____ (1)

c) Which of these numbers is a **square** number? _____ (1)

d) Which of these numbers is a **prime** number? _____ (1)

4. The fractions on the left are **equivalent** to <u>one</u> fraction on the right.

Draw a straight line to link each one. (4)

$$\frac{10}{16} \qquad\qquad\qquad \frac{10}{15}$$

$$\frac{2}{3} \qquad\qquad\qquad \frac{1}{2}$$

$$\frac{5}{10} \qquad\qquad\qquad \frac{9}{12}$$

$$\frac{6}{8} \qquad\qquad\qquad \frac{5}{8}$$

5.

a) Write the meaning of > and <: (1)

> _____

< _____

b) Use one of them and write it in the box to show which of these pairs of fractions is smaller. (2)

$$\frac{5}{9} \quad \boxed{} \quad \frac{2}{3}$$

$$\frac{4}{8} \quad \boxed{} \quad \frac{3}{7}$$

6.

a) Write all the factors of 10. (1)

_____

b) Write all the factors of 25. (1)

_____

c) What is the HCF of 10 and 25? (1)

_____

d) Write $\dfrac{10}{25}$ in its **simplest** form. (1)

_____

7.

a) Write the first six multiples of 8. (1)

_____

b) Write the first eight multiples of 6. (1)

_____

c) What is the LCM of 8 and 6? (1)

_____

d) Put these fractions in order starting with the largest.

(2)

$$\frac{3}{8} \qquad \frac{5}{6} \qquad \frac{7}{12}$$

_____

8. For my Birthday, I was given a box of 24 chocolates.

a) On the first day, I ate 6 chocolates.

What fraction of the box of chocolates did I eat?
Simplify your answer. (1)

_____

b) How many chocolates will there be in 18 boxes of chocolates?

(3)

_____

c) How many boxes contain a total of 216 chocolates? (2)

_____

9) Advance School has 521 children. One day 123 children went on a school trip. The rest of the children went to school. How many children went to school? (2)

_____

10) a) Write $\dfrac{14}{3}$ as a mixed number (1)

b) Write $3\dfrac{2}{7}$ as an improper (top heavy) fraction. (1)

Section 2
Topic 1 : Adding and Subtracting Fractions

OBJECTIVES

In this module, you will be able to:

❖ Find a **common denominator** for two fractions by finding the **Lowest Common Multiple (LCM).**

❖ Be able to add and subtract fractions with the same **denominator**

KEYWORDS:

Lowest Common Multiple   This is the smallest number that appears as a multiple of two numbers (the **LCM**).

Equivalent Fractions   Different fractions that represent the same decimal or whole number e.g. $\frac{1}{2} = \frac{2}{4} = 0.5$

Mixed Number   A number that has a whole number **and** a fraction e.g. $1\frac{2}{9}$, $5\frac{4}{5}$, $4\frac{9}{11}$

PREVIOUS KNOWLEDGE REQUIRED:

❖ Top Heavy Fractions
❖ Lowest Common Multiple
❖ Highest Common Factor
❖ Equivalent Fractions

Adding Wholes and Fractions

<u>Example 1</u>:   Find $1 + \frac{1}{5}$ and write your answer as an improper fraction.

<u>Answer</u>:   Notice that we can rewrite 1 as $\frac{5}{5}$ to suit our question:

$$\frac{5}{5} + \frac{1}{5}$$

$$= \frac{6}{5}$$

<u>Example 2</u>:   Find $1 - \frac{1}{5}$ and write your answer as an improper fraction.

<u>Answer</u>:   Similarly to Example 1:

$$1 - \frac{1}{5}$$

$$= \frac{5}{5} - \frac{1}{5}$$

$$= \frac{4}{5}$$

YOU TRY:

1.   Find the answer to the following:

a) $1 + \frac{1}{10}$ e) $1 - \frac{7}{9}$

b) $1 + \frac{1}{2}$ f) $1 - \frac{4}{10}$

c) $\frac{1}{5} + 1$ g) $1 - \frac{1}{2}$

d) $\frac{1}{8} + 1$ h) $1 - \frac{2}{7}$

Addition and Subtraction of Fractions with the same denominator

Example 1:   Find $\frac{4}{10} + \frac{2}{10}$.

Answer: We have the same denominator for both fractions, so simply add the numerators together:

$$\frac{4}{10} + \frac{2}{10}$$

$$= \quad \frac{4+2}{10}$$

$$= \quad \frac{6}{10}$$

This fraction can be simplified or cancelled down to:

$$\frac{6}{10} = \frac{3}{5}$$

Example 2:   Work out $\frac{4}{10} - \frac{2}{10}$.

$$= \quad \frac{4-2}{10}$$

$$= \quad \frac{2}{10}$$

$$= \quad \frac{1}{5}$$

CAUTION:

When we add or subtract fractions together, we <u>DO NOT</u> add or subtract the denominator (bottom number).

$\frac{4}{10} + \frac{2}{10} = \frac{6}{20}$ is <u>WRONG</u>   AND $\frac{4}{10} - \frac{2}{10} = \frac{2}{0}$ is <u>WRONG</u>.

YOU TRY:

2. Work out the answers to these fractions by adding or subtracting.

   Simplify your answer when necessary.

a) $\frac{1}{6} + \frac{2}{6}$

b) $\frac{1}{3} + \frac{2}{3}$

c) $\frac{1}{9} + \frac{5}{9}$

d) $\frac{9}{15} - \frac{4}{15}$

e) $\frac{1}{4} + \frac{2}{4}$

f) $\frac{7}{9} - \frac{4}{9}$

g) $\frac{7}{10} + \frac{3}{10}$

h) $\frac{4}{13} + \frac{9}{13}$

i) $\frac{16}{21} - \frac{2}{21}$

j) $\frac{29}{39} - \frac{11}{39}$

k) $\frac{17}{10} + \frac{7}{10}$

Addition and Subtraction of Fractions with different denominators

Example 1: Find $\frac{1}{3} + \frac{3}{6}$

Answer: Notice that the denominator is not the same. We will use the <u>lowest common multiple</u> to find one. (If you have forgotten, revise the topic)

The lowest common multiple of 3 and 6 is 6.

Write the equivalent fraction:

$$\frac{1}{3} = \frac{2}{6}$$

Now we can simply add the numerators:

$$\frac{2}{6} + \frac{3}{6}$$

$$= \frac{2+3}{6}$$

$$= \frac{5}{6}$$

<u>Example 2</u>: Find $\frac{4}{10} - \frac{1}{5}$.

<u>Answer</u>: $\frac{4}{10} - \frac{1}{5}$

$$= \frac{4}{10} - \frac{2}{10}$$

$$= \frac{4-2}{10}$$

$$= \frac{2}{10} = \frac{1}{5}$$

YOU TRY:

3. Calculate the answers to the following.

a) $\frac{1}{6} + \frac{2}{3}$

b) $\frac{9}{10} - \frac{2}{5}$

c) $\frac{2}{3} + \frac{1}{4}$

d) $\frac{1}{5} + \frac{2}{7}$

e) $\frac{3}{8} - \frac{1}{6}$

f) $\frac{3}{4} - \frac{2}{5}$

g) $\frac{19}{20} - \frac{4}{5}$

h) $\frac{4}{5} - \frac{3}{10}$

i) $\frac{7}{10} - \frac{11}{20}$

j) $\frac{5}{8} + \frac{1}{24}$

k) $\frac{7}{8} - \frac{2}{3}$

l) $\frac{5}{6} + \frac{1}{12}$

m) $\frac{1}{5} + \frac{3}{10}$

n) $\frac{11}{12} - \frac{3}{4}$

4. Review and copy each question into your book and fill in the blanks with the correct numbers. The first one has been done for you:

a) $\frac{1}{3} + \frac{?}{3} = 1$ **Since 1 is $\frac{3}{3}$ We need 2 more. So ? is 2.**

b) $\frac{1}{6} + \frac{?}{6} = \frac{5}{6}$

h) $1 - \frac{2}{5} = \frac{?}{?}$

i) $1 - \frac{6}{8} = \frac{?}{?}$

c) $\frac{1}{5} + \frac{?}{5} = \frac{3}{5}$

j) $\frac{7}{10} + \frac{?}{10} = \frac{18}{20}$

d) $\frac{1}{2} + \frac{?}{5} = \frac{9}{10}$

k) $\frac{1}{2} + \frac{?}{8} = \frac{5}{8}$

e) $\frac{3}{8} + \frac{?}{8} = 1$

l) $\frac{1}{3} + \frac{?}{9} = \frac{5}{9}$

f) $\frac{5}{6} + \frac{?}{6} = 1$

m) $\frac{2}{6} + \frac{?}{3} - \frac{4}{6}$

g) $1 - \frac{3}{6} = \frac{?}{?}$

5. **Work out the answers to these fractions by adding or subtracting.**

a) $\frac{1}{3} + \frac{1}{2} =$

g) $\frac{1}{5} + \frac{1}{10} =$

b) $\frac{1}{3} + \frac{1}{6} =$

h) $\frac{2}{5} + \frac{1}{10} =$

c) $\frac{1}{2} + \frac{1}{4} =$

i) $\frac{3}{5} + \frac{3}{10} =$

d) $\frac{1}{2} + \frac{3}{4} =$

j) $\frac{4}{5} + \frac{7}{10} =$

e) $\frac{2}{3} + \frac{1}{6} =$

k) $\frac{5}{8} - \frac{2}{9} =$

f) $\frac{1}{3} + \frac{4}{6} =$

l) $\frac{4}{10} - \frac{3}{12} =$

6. Fraction Problems

Solve the following problems using addition or subtraction.

a) Reena has $\frac{2}{5}$ m of pink ribbon and $\frac{3}{4}$ m of blue ribbon. She joins them together. What length of ribbon does she have now?

b) Ola had $\frac{7}{10}$ of her chocolate bar left. She gave $\frac{1}{3}$ away to her sister, Abi. What fraction of her chocolate is left?

c) Our cooking recipe asks for $\frac{1}{2}$ cup of sugar and $\frac{3}{8}$ cup of syrup. How much sugar and syrup is required altogether?

d) In the school garden, we planted $\frac{5}{12}$ packets of sunflower seeds and left $\frac{1}{3}$ of the packet. What fraction of the seeds were in the packet before we planted?

e) Maya has used $\frac{3}{5}$ of the space on her hard drive of her laptop. She wants to download a software that will use up another $\frac{1}{7}$ of the hard drive space. What fraction of her drive is left?

Section 2
Topic 2 : Fractions of Numbers

OBJECTIVES

In this module, you will be able to:

❖ Use <u>multiplication</u> and <u>division</u> to find **fractions of numbers**

KEYWORDS:

Fraction

A number consisting of a top number called the numerator and a bottom number called the denominator e.g. $\frac{4}{10}, \frac{5}{8}$.

Previous knowledge:

❖ Multiplication and Division

Fractions of numbers – when the numerator is 1

<u>Example 1:</u>   Find $\frac{1}{5}$ of 50.

<u>Answer:</u>   The word 'of ' in the question means to <u>multiply</u>:

$\frac{1}{5}$ x 50

You may want to write 50 as $\frac{50}{1}$
(because 50 divided by 1 is still 50)

$\frac{1}{5}$ x $\frac{50}{1}$

Multiplying:

$=$ $\frac{50}{5}$
$=$ <u>10</u>

YOU TRY:

1.   Find the fractions of the quantities.

a) $\frac{1}{5}$ of 20

b) $\frac{1}{2}$ of 10

c) $\frac{1}{4}$ of 100

d) $\frac{1}{9}$ of 36

e) $\frac{1}{7}$ of 21

f) $\frac{1}{7}$ of 28

g) $\frac{1}{3}$ of 33

h) $\frac{1}{2}$ of 30

i) $\frac{1}{4}$ of 20

j) $\frac{1}{5}$ of 15

k) $\frac{1}{6}$ of 36

l) $\frac{1}{6}$ of €42

m) $\frac{1}{8}$ of £16

n) $\frac{1}{4}$ of $4

o) $\frac{1}{8}$ of €48

p) $\frac{1}{2}$ of €8

q) $\frac{1}{7}$ of $14

r) $\frac{1}{6}$ of £18

s) $\frac{1}{11}$ of $44

t) $\frac{1}{14}$ of $28

u) $\frac{1}{20}$ of £120

Fractions of numbers – when the numerator is not 1

Example: Find $\frac{2}{5}$ of 50.

Method 1

Using the exact same method as before:

$$\frac{2}{5} \times 50$$

$$= \quad \frac{2}{5} \times \frac{50}{1}$$

$$= \quad \frac{2}{1} \times \frac{10}{1}$$

$$= \quad \frac{20}{1}$$

$$= \quad \underline{20}$$

Method 2

We know that for the fraction $\frac{2}{5}$, it means 2 <u>divided</u> by 5.

We can therefore divide 50 by 5 first:

$$50 \div 5$$

$$= \quad 10$$

Now multiply our answer by 2:

$$10 \times 2$$

$$= \quad \underline{20}$$

YOU TRY:

2.   Find the fractions of the quantities.
        The first one has been done for you.

a) $\frac{3}{5}$ of 20 = 20 ÷ 5 = 4 x 3 = 12.

s) $\frac{10}{11}$ of 44 degrees C

b) $\frac{2}{5}$ of 20

t) $\frac{11}{14}$ of 28 lb

c) $\frac{3}{4}$ of 80

u) $\frac{17}{20}$ of 20 inches

d) $\frac{2}{9}$ of 27

e) $\frac{2}{7}$ of 28

f) $\frac{3}{7}$ of 14

g) $\frac{2}{3}$ of 33

h) $\frac{2}{3}$ of 12

i) $\frac{2}{4}$ of 20

j) $\frac{4}{5}$ of 25

k) $\frac{5}{6}$ of 42

l) $\frac{4}{6}$ of 36 meters

m) $\frac{3}{8}$ of 16 grams

n) $\frac{3}{4}$ of 4 ml

o) $\frac{2}{8}$ of 48 pence

p) $\frac{3}{4}$ of 8 mm

q) $\frac{5}{7}$ of 14 kg

r) $\frac{3}{8}$ of 24 miles

3. A piece of wood is 2 metre long. It is cut into halves.
How many metres long is each half?

4.

    a) How many letters are in the name **Borehamwood**?

    b) How many of them are the letter 'o'?

    c) Change this number to a fraction of the number of letters in **Borehamwood**.

5.

    a) How many letters are in the name **Hilltop**?

    b) How many of them are the letter 'l'?

    c) Change this number to a fraction of the number of letters in **Hilltop.**

6. On a bookshelf there are 24 books. $\frac{1}{2}$ are sold.

    a) How many are left?

    b) What is a fifth of 25?

7. Sweets cost 36p.

    a) What is a half of this?

    b) What is a quarter of this?

8.

    a) How many letters are in the alphabet?

    b) How many letters are in half of the alphabet?

    c) How many letters are in a $\frac{7}{26}$ of the alphabet?

9.

    a) How many minutes are in an hour?

    b) How many minutes in a $\frac{1}{4}$ of an hour?

    c) How many minutes in $\frac{3}{4}$ of an hour?

Section 2
Topic 3 : Decimals

OBJECTIVES

In this module, you will learn:

❖ What a decimal is

❖ What tenths, hundredths and thousandths are

❖ How to order decimals

KEYWORDS:

Decimal	A number which has a decimal point in it e.g. 2.5, 1.8, 5.78, 11.497
Tenths	The first digit after the decimal point
Hundredths	The second digit after the decimal point
Thousandths	The third digit after the decimal point

PREVIOUS KNOWLEDGE:

❖ Figures

❖ Place Values

Decimals

Consider the following numbers:

2.5   1.8   55.76

The dot in these numbers is known as the **decimal point**.

We call these types of numbers **decimals**.

2.5   is read as "two point five".

1.8   is read as "one point eight".

55.76 is read as "fifty five point seven six".

Note that the numbers after the decimal point are read as their digits
*E.g. "point seven six" **NOT** "point seventy six".*

Decimals – Tenths

➢   Recapping from previous weeks, we know that the number 15 has 1 ten and 5 units.

➢   We can do the same for decimals.

➢   The first number after the decimal point are known as **tenths**.

          2.5   has 2 units and 5 **tenths**.

          1.8   has 1 unit and 8 **tenths**.

          55.76 has 5 units and 7 **tenths**.

➢   We can split these numbers up to see how these numbers are actually made up.

          2.5   = 2.0 + 0.5   The 0.5 represents 5 tenths.

          1.8   = 1.0 + 0.8   The 0.8 represents 8 tenths.

          5.76 = 5.00 + 0.70 + 0.06   The 0.70 represents 7 tenths.

YOU TRY:

1. Copy each question and write how many tenths there are in each one.
   The first one has been done for you.

   a) 3.5 has <u>5 tenths</u>.

   b) 5.2 has __ tenths.   **i)** 1.447 has __ tenths.

   c) 9.4 has __ tenths.   **j)** 2.64 has __ tenths.

   d) 4.6 has __ tenths.   **k)** 4.105 has __ tenths.

   e) 3.1 has __ tenths.   **l)** 8.213 has __ tenths.

   f) 2.6 has __ tenths.   **m)** 7.331 has __ tenths.

   g) 9.2 has __ tenths.   **n)** 5.54 has __ tenths.

   h) 4.2 has __ tenths.   **o)** 2.350 has __ tenths.

2. Write in words the following decimals.
   The first one has been done for you.

   a) 0.2 is "<u>nought point two</u>".

   b) 1.6

   c) 22.8

   d) 7.3

   e) 9.3

   f) 3.4

   g) 4.97

   h) 8.41

   i) 2.582

Decimals – Hundredths

➢ The second number after the decimal point is known as **hundredths**.

2.74 has 4 hundredths.

1.35 has 5 hundredths.

16.92 has 2 hundredths.

➢ Again, we can split the numbers up to see how they are made up:

2.74 = 2.00 + 0.70 + 0.04   The 0.04 is the 4 hundredths.

1.35 = 1.00 + 0.30 + 0.05   The 0.05 is the 5 hundredths.

16.92 = 16.00 + 0.90 + 0.02   The 0.02 is the 2 hundredths.

YOU TRY:

3. Copy the sentences and fill in the missing numbers.

a) 3 units, 1 tenth and 5 hundredths is 3.15.

b) 8 units, 9 tenths and 3 hundredths is ____.

c) 4 units, 2 tenths and 7 hundredths is ____.

d) 9 units, 2 tenths and 1 hundredth is ____.

e) __ units, 8 tenths and 2 hundredths is 3.82.

f) 8 units, __ tenths and 5 hundredths is 8.35.

g) 7 units, 2 tenths and __ hundredths is 7.29.

h) 1 unit, 4 tenths and __ hundredths is 1.47.

i) 4 units, __ tenths and 5 hundredths is 4.55.

Decimals – Thousandths

➤ The third number after the decimal point is known as **thousandths**.

1.526 has <u>6 thousandths</u>.

2.690 has <u>0 thousandths</u>.

32.1294 has <u>9 thousandths</u>.

1.526 = 1 + 0.5 + 0.02 + <u>0.006</u>   0.006 are the thousandths.

2.690 = 2 + 0.6 + 0.09 + <u>0.000</u>   No thousandths.

7.129 = 7 + 0.1 + 0.02 + <u>0.009</u>   0.009 are the thousandths.

YOU TRY:

4. Copy each question and write the correct decimal for the worded numbers. The first one has been done for you.

a) 7 thousandths is <u>0.007</u>.

b) 8 thousandths is _____.

c) 2 thousandths is _____.

d) 5 thousandths is _____.

e) 1 thousandth is _____.

f) 6 thousandths is _____.

g) 9 hundredths is _____.

h) 1 tenth is _____.

i) 9 tenths is _____.

j) 7 hundredths is _____.

k) 5 tenths is _____.

l) 3 hundredths is _____.

m) 8 tenths is _____.

n) 6 thousandths is _____.

o) 6 hundredths is _____.

5. Copy the table into your book.
For the following numbers, write out how many tenths, hundredths and thousandths there are.

Number	Tenths	Hundredths	Thousandths
7.253	2	5	3
2.630			
22.984			
1.401			
8.014			
0.3061			
3.15			
9.251			
6.543			
1.1231			
3.0728			
82.5182			
9.2016			

Ordering Decimals

Consider the number:

$$1234.567$$

Reading from left to right, it has:

- 1 thousand

- 2 hundreds

- 3 tens

- 4 units

- 5 tenths

- hundredths

- thousandths

We know:

- Any number of thousands is **larger** than hundreds.

- Any number of hundreds is **greater** than tens.

- Any number of tens is **bigger** than units.

It follows on that:

- Any number of units is <u>larger</u> than tenths.

- Any number of tenths is <u>greater</u> than hundredths.

- Any number of hundredths is <u>bigger</u> than thousandths.

Of course, we don't include the number 0 in this explanation.

Example: Order the following decimals starting with the largest.

0.12   0.103   0.212   0.099

Steps

1.   From left to right, look at each digit one at a time.
       Starting with the units:

0.12   0.103   0.212   0.099

They are all 0, so let's move onto the next digit.

2.   Looking at tenths:

0.12   0.103   0.212   0.099

0.12 and 0.103 have 1 tenth. Save these for Step 3.

0.212 has 2 tenths.

0.099 has 0 tenths.

We conclude that 0.212 is largest and 0.099 is smallest.

3.   Now let's look at the two numbers with 1 tenth:

0.12   0.103

0.12   has 2 hundredths.
0.103 has 0 hundredths.

0.12 is <u>larger</u> than 0.102.

4.   The order starting with the biggest is:

0.212   0.12   0.103   0.099

YOU TRY:

6.   Order the sets of decimals in each question starting with the largest.

a)   0.72   0.13   0.28   0.27

b)   0.28   1.30   0.30   0.55

c)   0.52   0.55   0.45   0.43

d)   4.32   2.95   4.21   4.12

e)   0.02   0.84   0.13   0.15

f)   1.49   2.30   0.13   1.34

g)   1.231   1.291   1.294   1.295

7.   Order the sets of decimals in each question starting with the smallest.

a)   9.902   9.901   9.909   9.755

b)   1.32   1.33   1.38   1.39

c)   5.20   5.32   5.39   5.297

d)   2.532   2.913   3.194   3.200

e)   1.488   1.391   1.199   1.481

f)   0.110   0.119   0.111   0.117

g)   53.14   34.21   34.23   53.11

h)   8.19   8.13   8.109   7.999

Section 2
Topic 4 : Decimal and Fraction Conversions

OBJECTIVES

In this module, you will:

❖ Revise about tenths and hundredths

❖ Understand how to convert decimals into fractions

❖ Be able to convert fractions into decimals

KEYWORDS:

Decimal

A number which has a decimal point in it
e.g. 2.5, 1.8, 5.7, 11.4

Tenths

The first digit after the decimal point
In fractions, the denominator is 10

Hundredths

The second digit after the decimal point
In fractions, the denominator is 100

PREVIOUS KNOWLEDGE:

❖ Decimals
❖ Equivalent Fractions
❖ Mixed Numbers

Decimal Tenths and Fractions

In previous weeks, we learnt that $\frac{5}{8}$ is read as five <u>eighths</u>.

We know the first number after the decimal point is known as <u>tenths</u>.

Therefore:

<u>Fraction</u>	<u>Words</u>	<u>Decimal</u>
$\frac{5}{10}$	Five tenths	0.5
$\frac{8}{10}$	Eight tenths	0.8
$\frac{7}{10}$	Seven tenths	0.7

The denominator will <u>always</u> be <u>10</u> for <u>tenths</u>.

YOU TRY:

1. Copy and fill in the blank spaces appropriately.
    The first one has been done for you.

a) 0.4 has 4 tenths.

b) 0.6 has __ tenths.

c) 0.7 has __ tenths.

d) 0.3 has __ tenths.

e) 0.8 has __ tenths.

f) $\frac{1}{10}$ is the same as 0.1

g) $\frac{1}{10}$ is the same as ___.

h) $\frac{3}{10}$ is the same as ___.

I) $\frac{2}{10}$ is the same as ___.

j) $\frac{4}{10}$ is the same as ___.

Decimal Hundredths and Fractions

As a reminder, the second number after the decimal point are known as **hundredths**.

Fraction	Words	Decimal
$\dfrac{4}{100}$	Four hundredths	0.04
$\dfrac{5}{100}$	Five hundredths	0.05
$\dfrac{2}{100}$	Two hundredths	0.02

The denominator will <u>always</u> be 100 for <u>hundredths</u>.

YOU TRY:

2.   Copy and complete the table.

Fraction	Words	Decimal
$\dfrac{7}{100}$		0.07
	Nine hundredths	
$\dfrac{1}{100}$		
		0.03

3. Copy and fill in the blanks.

a) 3.28 has __ tenths and __ hundredths.

b) 82.91 has __ tenths and __ hundredths.

c) 9.34 has __ tenths and __ hundredths.

d) 0.13 has __ tenths and __ hundredths.

e) 3.76 has __ tenths and __ hundredths.

f) 15.02 has __ tenths and __ hundredths.

g) 3.423 has __ tenths and __ hundredths.

h) $\frac{3}{100}$ as a decimal is _____.

i) $\frac{5}{100}$ as a decimal is _____.

j) $\frac{8}{10}$ as a decimal is _____.

k) $\frac{6}{10}$ as a decimal is _____.

l) $\frac{62}{100}$ as a decimal is _____.

m) $\frac{12}{100}$ as a decimal is _____.

n) $\frac{3}{10}$ as a decimal is _____.

Changing Fractions to Decimals

Let's look at this table:

Fraction	Words	Decimal
$\dfrac{5}{10}$	Five tenths	0.5
$\dfrac{9}{100}$	Nine hundredths	0.09

Remember that: $\dfrac{5}{10}$ is 5 divided by 10.

$\dfrac{9}{100}$ is 9 divided by 100.

Taking the numerator of each one and writing the decimal point in, we get 5.0 and 9.0.

● Dividing by 10 moves the decimal point <u>one place</u> to the <u>left</u>:

$\div\ 10$

0 5.0

0 .5 0

$5.0 \quad \rightarrow \quad \underline{0.5}$
$\div 10$

This shows $\dfrac{5}{10}$ is equal to <u>0.5</u>.

- Dividing by 100 moves the decimal point <u>two places</u> to the <u>left</u>.

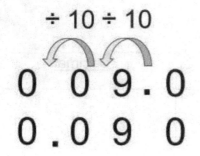

$$0\ \ 0\ \ 9.0$$
$$0.0\ \ 9\ \ 0$$

9.0   →   0.9   →   <u>0.09</u>
   ÷10   ÷10

This shows $\frac{9}{100}$ is equal to <u>0.09</u>.

Using this technique, we can work out decimals for more complicated fractions:

Fraction	Words	Decimal
$\frac{52}{100}$	Fifty two hundredths	0.52
$\frac{24}{100}$	Twenty four hundredths	0.24

<u>Question:</u>   If we multiply by 10 or 100 how does the decimal place move?

YOU TRY:

4. Convert these fractions and worded fractions into decimals.

a) $\frac{1}{10}$

b) $\frac{5}{10}$

c) $\frac{12}{100}$

d) $\frac{27}{100}$

e) $\frac{19}{100}$

f) $\frac{45}{100}$

g) $\frac{95}{100}$

h) $\frac{25}{10}$

i) $\frac{206}{100}$

j) $\frac{24}{10}$

k) $\frac{310}{100}$

l) Seventy one hundredths

m) Six hundredths

n) Four tenths

o) Thirty three hundredths

p) Five tenths and two hundredths

q) Twenty five hundredths

r) Twelve hundredths

s) One tenth

Changing Decimals to Fractions

<u>Example</u>:   Write 0.32 as a fraction.

<u>Answer</u>:

We can split this up into tenths and hundredths:

0.32   =   0.30   +   0.02

We know:

0.30   =   3 tenths   = $\frac{3}{10}$

0.02   =   2 hundredths = $\frac{2}{100}$

Therefore:

0.32   =   0.30   +   0.02

          = $\frac{3}{10}$ + $\frac{2}{100}$

          = $\frac{30}{100}$ + $\frac{2}{100}$

          = $\frac{32}{100}$

          = $\frac{16}{50}$

          = $\frac{8}{25}$

<u>Note</u>: if the number has units, tens etc. we get a <u>mixed number</u> instead.

<u>Example</u>:   Write 2.5 as a mixed number.

<u>Answer</u>: $2.5 \quad = \quad 2.0 \quad + \quad 0.5$

$= \quad 2.0 \quad + \quad \dfrac{5}{10}$

$= \quad 2\dfrac{5}{10}$

$= \quad 2\dfrac{1}{2}$

YOU TRY:

5.   Convert these decimals to fractions.

Simplify your answer when you can.

a)	0.2		**i)**	0.03
b)	0.5		**j)**	0.04
c)	0.75		**k)**	1.7
d)	0.1		**l)**	9.25
e)	0.3		**m)**	2.4
f)	0.8		**n)**	4.5
g)	0.6		**o)**	0.75
h)	0.7		**p)**	0.86

Section 2
Topic 5 : Adding and Subtracting Decimals

OBJECTIVES

In this module, you will:

❖ Revise the carry and borrowing rule in addition and subtraction

❖ Be able to line up decimals to add or subtract appropriately

❖ Apply decimals to money problems as well as others

KEYWORDS:

Decimal   A number which has a decimal point in it
e.g. 2.5, 1.8, 5.7, 11.4

Tenths   The first digit after the decimal point

Hundredths   The second digit after the decimal point

Thousandths   The third digit after the decimal point

PREVIOUS KNOWLEDGE:

❖ Digits

❖ Decimals

❖ Addition

❖ Subtraction

Adding Decimals

To add decimals, we must line up the tenths, hundredths etc.

It is exactly the same as lining up units, tens etc. for whole numbers.

<u>Important:</u>   We must line up the decimal point correctly.

This is shown in the following examples.

<u>Example 1</u>

Find 1·2 + 0·1.

<u>Example 2</u>

Add 1·02 to 2·09.

<u>Answer</u>

```
   1 · 2
+  0 · 1
   1 · 3
```
↑ Notice how the
   decimal points align

<u>Answer</u>

```
        1
   1 · 0 2
+  2 · 0 9
   3 · 1 1
```

2 + 9 gives us 11 hundredths.

We carry the 1 to the tenth spot.

We can use decimals to also help us find totals of money.

<u>Example 3</u>

Find the total of 40p and £2.50.

<u>Answer:</u>   40p is the same as £0·40.

```
   0 · 4 0
+  2 · 5 0
£  2 · 9 0
```
↑ Don't forget your units!

Exercise

1. Practise lining up the decimals to add and find the answer.

a)   1.1 + 0.8   g)   0.38 + 2.51

b)   1.6 + 0.2   h)   0.76 + 5.12

c)   1.4 + 4.3   i)   22.3 + 1.4

d)   1.3 + 3.5   j)   1.5 + 2.35

e)   7.23 + 2.11   k)   0.038 + 0.12

f)   6.17 + 0.71   l)   3.9 + 12.03

2. In this exercise, you will need to use the carrying method to
    find these sums. (Notice how the decimals are lined up with each
    other). Copy and complete.

a)
```
    6 · 9 6
  + 0 · 1 1
```

b)
```
   1 3 · 0 6
  +  7 · 1 9
```

c)
```
    4 · 2 0 9
  + 7 · 8 1
```

d)
```
   5 5 · 7
  +  4 · 6 8
```

e)
```
    6 · 9 0 7
  +2 0 · 8 1 4
```

f)
```
    7 · 3 6
  +1 2 · 2 4
```

g)
```
   4 6 · 0 6 4
  +  2 · 1 9 7
```

h)
```
   7 1 · 0 5
  +  4 · 9 5 1
```

3. Find the totals of the prices in the following questions.
    Convert them appropriately.

a)   28p and £5.12   c)   183p and £11.24

b)   59p and £3.17.   d)   £6.35 and 76p.

Subtracting Decimals

We can subtract decimals in the same way.

Example 1

Subtract 6.7 from 9.2.

Answer

```
  9 · 9
- 6 · 7
  3 · 2
```

Example 2

Subtract 5.15 from 8.9.

Answer

Write 8.9 as 8.90.

This is important as it shows we get 10 hundredths <u>after borrowing</u> 1 tenth.

```
8 · ⁸9̶ ¹0
- 5 · 1 5
  3 · 7 5
```

Example 3

I want to buy some rope costing 99p.

How much change will I get from a £5 note?

Answer:

99p is equivalent to £0.99.

We must subtract £0.99 from £5.

Therefore, the £5 **must** be on the <u>top</u>.

1)
```
  5 · 0 0
- 0 · 9 9
```

We have lined up the decimal points of the two numbers together.

2)
```
⁴5̶ · ⁹0̶ ¹0
-  0 · 9 9
```

In this step, we must borrow twice, one from the tenths column and one from the units column.

3)
```
⁴5̶ · ⁹0̶ ¹0
-  0 · 9 9
   4 · 0 1
```

The change is <u>£4.01</u>.

Exercise

4. For the following, practise lining up the decimals to subtract, and find the answer.

    a)   5.1 - 2.0

    b)   7.9 - 4.3

    c)   19.3 - 11.2

    d)   1.9 - 0.5

    e)   2.1 - 1.2

    f)   33.7 - 22.5

    g)   2.27 - 0.14

    h)   11.37 - 10.14

    i)   32.6 - 1.5

    j)   2.67 - 0.1

    k)   6.318 - 4.11

    l)   12.454 - 2.13

5. In this exercise, you will need to use the borrowing method twice to find these subtractions. Copy and complete.

    a) $\begin{array}{r} 6 \cdot 9\,4 \\ -\ 0 \cdot 7\,5 \\ \hline \end{array}$

    b) $\begin{array}{r} 1\,2 \cdot 4\,6 \\ -\ \ 3 \cdot 1\,7 \\ \hline \end{array}$

    c) $\begin{array}{r} 9 \cdot 1\,4 \\ -\ 8 \cdot 0\,2\,5 \\ \hline \end{array}$

    d) $\begin{array}{r} 1\,3 \cdot 6 \\ -\ \ 5 \cdot 5\,4 \\ \hline \end{array}$

    e) $\begin{array}{r} 6 \cdot 2\,3\,4 \\ -\ 2 \cdot 1\,4 \\ \hline \end{array}$

    f) $\begin{array}{r} 7 \cdot 4\,0\,5 \\ -\ 3 \cdot 2\,4\,4 \\ \hline \end{array}$

    g) $\begin{array}{r} 4\,6 \cdot 4 \\ -\ \ 2 \cdot 1\,1\,1 \\ \hline \end{array}$

    h) $\begin{array}{r} 6\,6 \cdot 0\,5 \\ -\ \ 4 \cdot 9\,4\,1 \\ \hline \end{array}$

6.

a) I buy a cookie costing 60p.

How much change do I get from a £2 coin?

b) I buy a lamp costing £3.99.

How much change do I get from a £5 note?

c) A thermometer costs £2.65.

You pay with a £5 note.

What is the change?

d) A calculator costs £7.50.

You pay with a £10 note.

What is the change?

e) It costs me £4.70 to post a parcel.

What change do I get if I pay with a £5 note?

f) Chocolate muffins cost £1.25.

Blueberry muffins cost 90p.

What is the difference in price?

g) The hardback of a book costs £12.40.

The paperback costs £10.40.

How much more does the hardback book cost than the paperback?

Section 2
Topic 6 : Multiplying and Dividing Decimals

OBJECTIVES

In this module, you will:

❖ Revise multiplying and dividing by 10s

❖ Be able to multiply whole numbers with decimals

❖ Be able to find the cost of one item in many items

KEYWORDS:

Decimal   A number which has a decimal point in it
                         e.g. 2.5, 1.8, 5.7, 11.4

Decimal Places   Refer to how many numbers there are after
                         the decimal point

Examples

a)   2.5 has <u>one</u> decimal place as there is only <u>one</u> digit <u>after</u> the
        decimal point

b)   3.62 has <u>two</u> decimal places as there is only <u>two</u> digits <u>after</u>
        the decimal point

PREVIOUS KNOWLEDGE:

❖ Digits
❖ Decimals
❖ Multiplication
❖ Division

Multiplying by 10s

x 10

Multiplying by 10 moves the decimal point <u>one place</u> to the <u>right</u>:

0. 4 0
0 4 .0

$$0.4 \quad \xrightarrow{\times 10} \quad 4.0$$

● Multiplying by 100 moves the decimal point <u>two places</u> to the <u>right</u>:

x 10 x 10

0. 0 0 8
0 0 0 .8

$$0.008 \quad \xrightarrow{\times 10} \quad 0.08 \quad \xrightarrow{\times 10} \quad 0.8$$

Multiplying by 1000 moves the decimal point <u>three places</u> to the <u>right</u>:

x10 x10 x10

0. 0 0 7 0
0 0 0 7 .0

$$0.007 \quad \xrightarrow{\times 10} \quad 0.07 \quad \xrightarrow{\times 10} \quad 0.7 \quad \xrightarrow{\times 10} \quad 7.0$$

<u>Examples</u>

a)   0.5 x 10 = <u>5</u>

b)   0.07 x 100 = <u>7</u>

c)   0.29 x 100 = <u>29</u>

d)   4.012 x 1000 = <u>4012</u>

You Try

1.   Copy and find the answers to the following multiplications:

      a)   3.2 x 10

      b)   6.3 x 10

      c)   5.2 x 10

      d)   7.9 x 10

      e)   2.6 x 10

      f)   0.64 x 100

      g)   0.35 x 100

      h)   1.52 x 100

      i)   3.73 x 100

      j)   8.21 x 100

      k)   6.313 x 1000

      l)   7.052 x 1000

      m)   0.002 x 1000

      n)   0.251 x 1000

      o)   2.672 x 1000

p) One can of sweet corn costs £0.80. 10 cans cost £_____.	**q)** One tub of yoghurt weighs 50g.  100 tubs weigh _____g.

Dividing by 10s

- Dividing by 10 moves the decimal point <u>one place</u> to the <u>left</u>:

$$\overset{\div 10}{\overset{\frown}{1\ 0.2}}$$

1 . 0 2

$$10.2 \quad \rightarrow \quad \underline{1.02}$$
$$\div 10$$

- Dividing by 100 moves the decimal point <u>two places</u> to the <u>left</u>:

$$\overset{\div 10\ \ \div 10}{\overset{\frown\frown}{1\ 2\ 3.4}}$$

1 . 2 3 4

$$123.4 \quad \rightarrow \quad 12.34 \rightarrow \quad \underline{1.234}$$
$$\div 10 \qquad\quad \div 10$$

- Dividing by 1000 moves the decimal point <u>three places</u> to the <u>left</u>.

$$\overset{\div 10\ \div 10\ \div 10}{\overset{\frown\frown\frown}{0\ \ 0\ \ 0\ 5.0}}$$

0 . 0 0 5 0

$$5.0 \quad \rightarrow \quad 0.5 \quad \rightarrow \quad 0.05 \quad \rightarrow \quad \underline{0.005}$$
$$\div 10 \qquad\quad \div 10 \qquad\quad \div 10$$

<u>Examples</u>

a) $5.7 \div 10 = \underline{0.57}$

b) $123.4 \div 100 = \underline{1.234}$

c) $29 \div 100 = 29.0 \div 100 = 0.290 = \underline{0.29}$

d) $4012 \div 1000 = 4012.0 \div 1000 = 4.0120 = \underline{4.012}$

Note: Remember to write the decimal point yourself if it is not there already.

You Try

2.   Copy and find the answers to the following divisions:

a)   7.3 ÷ 10

b)   1.24 ÷ 10

c)   67.2 ÷ 10

d)   9.0 ÷ 10

e)   28.4 ÷ 10

f)   25.2 ÷ 100

g)   32.1 ÷ 100

h)   100.2 ÷ 100

i)   275.3 ÷ 100

j)   1.24 ÷ 100

k)   674.8 ÷ 1000

l)   700.4 ÷ 1000

m)   110.4 ÷ 1000

n)   28.324 ÷ 1000

o)   3959 ÷ 1000

p) Ten cartons of milk costs £12.50.  One carton costs £_____ .	q) One hundred sweet bags costs £102.00. One sweet bag costs £_____ .

Multiplying decimals with whole numbers

<u>Example 1</u>

Let's try to multiply these two numbers:

$$1.2 \times 3$$

We do this <u>differently</u> compared to adding and subtracting.

<u>Step 1</u>

Ignore the decimal point in the 1.2

Treat it as the number 12.

<u>Step 2</u>

Now multiply the numbers 12 and 3 together in the usual way:

```
      1 2
  x     3
  ───────
      3 6
```

<u>Step 3</u>

1.2 has <u>one decimal place</u> because there is <u>one digit</u> after the decimal point.

Our answer must also have <u>one decimal place</u> in it.

Putting the decimal point back into our answer, we get <u>3.6</u>.

Example 2

Multiply 4 with 0.12.

Step 1

Ignore the decimal point in the 0.12 and treat it to be 12.

Step 2

Now multiply the numbers 12 and 4 together the usual way.

Remember the carry method.

We carry the 1 from 2 x 5 to the tens spot.

$$
\begin{array}{r}
{}^{1} \\
1\,2 \\
\times\quad 5 \\
\hline
5\,0
\end{array}
$$

Step 3

0.12 had <u>two decimal places</u> because there are <u>two digits</u> after the decimal point.

Our answer must also have <u>two decimal places</u> in it.

Putting the decimal point back into our answer, we get <u>0.48</u>.

You Try

3.   Find the answers to the following multiplications.

a) 1.3 x 2		**e)** 6.2 x 3		**i)** 2.3 x 7	
b) 2.3 x 3		**f)** 4.1 x 5		**j)** 3.1 x 5	
c) 1.1 x 7		**g)** 0.3 x 8		**k)** 7.21 x 6	
d) 4.2 x 2		**h)** 0.7 x 6		**l)** 3.12 x 3	

Dividing decimals by whole numbers

<u>Example 1</u>

Look at this division:

$$1 \cdot 2 \div 3$$

<u>Step 1</u>

As you are used to, write the calculation for the division.

$$3 \overline{\smash{\big)}\ 1 \cdot 2}$$

<u>Step 2</u>

Write the decimal point on the top <u>in line</u> with the one below:

$$\overset{\textstyle \cdot}{3 \overline{\smash{\big)}\ 1 \cdot 2}}$$

<u>Step 3</u>

3 divides 1 zero times.

We carry the 1 to the next digit to the right of it to make 12.

3 divides 12 four times.

$$\overset{\textstyle 0 \cdot 4}{3 \overline{\smash{\big)}\ 1 \cdot {}^{1}2}}$$

<u>Example 2</u>

$$6 \overline{\smash{\big)}\ 3 \cdot 06} \quad = \quad \overset{\textstyle 0 \cdot}{6 \overline{\smash{\big)}\ 3 \cdot {}^{3}06}} \quad = \quad \overset{\textstyle 0 \cdot 51}{6 \overline{\smash{\big)}\ 3 \cdot {}^{3}06}}$$

| 6 divides 3 zero times. Carry the 3 to make 30. | 6 divides 30 five times. 6 divides 6 one time. |

You Try

4. Divide the decimals by the whole number:

a)	1.4 ÷ 2	**e)**	6.9 ÷ 3	**i)**	2.5 ÷ 5
b)	3.6 ÷ 3	**f)**	3.6 ÷ 6	**j)**	6.4 ÷ 8
c)	1.5 ÷ 5	**g)**	9.6 ÷ 3	**k)**	6.24 ÷ 6
d)	2.2 ÷ 2	**h)**	7.6 ÷ 2	**l)**	3.96 ÷ 3

Money Problems

We can use our knowledge about decimals to help use solve money problems.

In every day life, people use <u>pence (p)</u> and <u>pounds (£)</u>.

To convert:

x 100

Pence (p)   Pounds (£)

÷ 100

<u>Example 1:</u>   Write 18p in pounds (£).

<u>Answer:</u>   We are looking at pence to pounds.

To convert, we divide the pence by 100:

18 ÷ 100 = 18.0 ÷ 100 = <u>£0.18</u>

<u>Example 2:</u>   Write £0.32 in pence (p).

<u>Answer:</u>   We are looking at pounds to pence this time.

To convert, we multiply the pounds by 100:

0.32 x 100 = <u>32p</u>

You Try

5.   Write the money in Pounds (£).

a)   60p

b)   78p

c)   93p

d)   25p

e)   205p

f)   312p

g)   44p

h)   69p

i)   709p

6.   Write the money in Pence (p).

a)   £2.50

b)   £1.40

c)   £0.29

d)   £4.65

e)   £2.12

f)   £5.10

g)   £9.05

h)   £3.90

i)   £60.74

7.
Copy the sentence and fill in the answer.

4
Quality
Pencils

80p

Each pencil costs _____.

30p

6 donuts costs _____.

£2.40

One book costs _____.

One apple costs 25p.

8 apples cost _____.

2 bottles £2.50

One bottle of coke costs _____.

£95.50

Two computers cost _____.

Multiplying decimals by another decimal

Example 1

$$3{\cdot}2 \times 2.1$$

Step 1

To make it easier to work out, you write it out again without the decimal points.

3.2 x 2.1 becomes 32 x 21

Step 2

To multiply 32 by 21, first multiply 32 by 1

```
      3 2
    × 2 1
      3 2  ←——— 3 2 x 1 = 3 2
    6 4 0  ←——— put a zero down and multiply 32 by 2
    6 7 2  ←——— Add 32 to 640
```

The answer is 6 7 2

Remember, any number can be changed to a decimal. 3 is the same as 3.0, 11 is the same as 11.0 and 672 is the same as 672.0

Step 3

The numbers we started with were 3.2 and 2.1
There are **2** numbers to the right of the decimal points. So, starting at the right hand side of 672, we jump back **2** places.

÷ 10 ÷ 10

6 7 2 . 0

6 . 7 2 0

6 7 2 . 0 → 6 . 7 2 0

which is the same as 6 . 7 2

That is, the answer is **6.72**

You Try

8. Find the answers to the following multiplications.

a)	2.3 x 2.1	**f)**	4.1 x 5.7	**k)**	2.1 x 6.3
b)	3.7 x 5.3	**g)**	6.3 x 3.5	**l)**	9.5 x 1.3
c)	1.1 x 7.7	**h)**	7.1 x 6.9		
d)	2.1 x 4.8	**i)**	3.3 x 7.1		
e)	5.2 x 3.1	**j)**	9.2 x 3.6		

Section 2
Topic 7 : Percentages

OBJECTIVES

In this module, you will:

❖ Understand what a percentage is

❖ Be able to find 25%, 50% or 75% of a value

❖ Be able to find 10% of a value

❖ Use 10% to find 20%, 30%, 40% etc.

❖ Be able to find percentage increases and decreases of a value

KEYWORDS:

Percent   Written as %.

Percentages   A way to express a number out of 100
                     e.g. 30%, 60%, 82%

Increase   To "go up" in value

Decrease   To "go down" in value

PREVIOUS KNOWLEDGE:

❖ Multiplication and Division
❖ Addition and Subtraction
❖ Decimals

Understanding percentages

You may have already seen percentages from your test marks.

If you get everything correct, your mark will be 100%.

They help us to <u>compare</u> values with one another.

Percent is written as %.

Percentages look like this:

50<u>%</u>

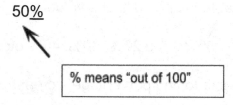

% means "out of 100"

This reads as "50 percent".

Finding 25%, 50% and 75%

In this module, we will look at questions where the given value will be 100%. This will be made clear in the following examples.

Note that:

100% ÷ 4 = <u>25%</u>

100% ÷ 2 = <u>50%</u>

100% ÷ 4 x 3 = <u>75%</u>

We will use this to help us find 25%, 50% and 75%.

Example 1:   Find 50% of 40.

Answer: The number 40 is 100%.

We know 100% ÷ **2** = 50%

So to find 50%, we divide 40 by **2**:

40 ÷ 2 = 20

50% of 40 is 20.

Example 2:   Find 25% of 28.

Answer: The number 28 is 100%.

We know 100% ÷ **4** = 25%

So to find 25%, we divide 28 by **4**:

28 ÷ 4 = 7

25% of 28 is 7.

Example 3:   Find 75% of 60.

Answer: The number 60 is 100%.

First find 25%: divide 60 by 4:   60 ÷ 4 = **15**

To find 75%, we now multiply by 3:   **15** x 3 = 45

75% of 60 is 45.

YOU TRY:

1. Copy and find 50% of the following numbers:

a)	50% of 16	**e)**	50% of 8	**h)**	50% of 24
b)	50% of 20	**f)**	50% of 14	**I)**	50% of 36
c)	50% of 26	**g)**	50% of 18	**j)**	50% of 46
d)	50% of 12				

k) A book at the library costs £32.

50% of the price is _____

l) Linna needs to drive 40 miles. She's only half way.

So far, she's driven _____ miles.

2. Copy and find 25% of the following numbers:

a)	25% of 28	**e)**	25% of 8	**h)**	25% of 48
b)	25% of 16	**f)**	25% of 64	**I)**	25% of 36
c)	25% of 32	**g)**	25% of 40	**j)**	25% of 44
d)	25% of 20				

k) Larry's test mark is 25% of the total 80.

He achieved _____ many marks.

l) Arfan needs to measure 25% of 100m.

25% of 100 is _____ m.

3. What is 75% of the following numbers?

Remember: First find 25% and then multiply your answer by 3.

a) 75% of 28 miles.   **e)** 75% of 12 km.

b) 75% of 20 hours.   **f)** 75% of £40.

c) 75% of 8g.   **g)** 75% of $24.

d) 75% of 16 litres.   **h)** 75% of 64°.

Finding 10%

Note: $100\% \div 10 = \underline{10\%}$
Similarly, to find 10%, we divide the number in the question by <u>10</u>.

Example: Find 10% of 40.

Answer: The number 40 is 100%.

We know $100\% \div \mathbf{10} = 10\%$

To find 10%, we divide 40 by **10**: $40 \div 10 = \underline{4}$

10% of 40 is <u>4</u>.

YOU TRY:

4. What is 10% of the following numbers?

a) 10% of 10   **d)** 10% of 100   **g)** 10% of 90

b) 10% of 20   **e)** 10% of 110   **h)** 10% of 120

c) 10% of 70   **f)** 10% of 40

Finding 20%, 30%, 40%....

Remember: 100% ÷ 10 = <u>10%</u>

Therefore, we can find 20%, 30%.... using multiples of 10% simply by multiplying by 2, 3 etc.

<u>Example 1</u>:   Find 20% of 40.

<u>Answer</u>:   The number 40 is 100%.

First find 10%: divide 40 by 10:   40 ÷ 10 = **4**

To find <u>2</u>0%, we now multiply by <u>2</u>:   **4** x 2 = <u>8</u>

20% of 40 is <u>8.</u>

<u>Example 2</u>:   What is 40% of 10?

<u>Answer</u>:   The number 10 is 100%.

First find 10%: divide 10 by 10:   10 ÷ 10 = **1**

To find <u>4</u>0%, we now multiply by <u>4</u>:   **1** x 4 = <u>4</u>

40% of 10 is <u>4.</u>

YOU TRY:

5.   What are the percentages of the following numbers?

a)	20% of 10	**f)**	80% of 40	**j)**	20% of 260
b)	30% of 30	**g)**	70% of 150	**k)**	30% of 370
c)	60% of 20	**h)**	20% of 230	**l)**	40% of 480
d)	20% of 100	**i)**	30% of 70	**m)**	60% of 250
e)	40% of 90				

Increases and Decreases

What do we mean by <u>increases</u> and <u>decreases</u>?

<u>Increases</u>:   To "go up" in value

<u>Decreases</u>:   To "go down" in value

In a shop, they <u>discount</u> items and we pay <u>less</u> for them.

This is a <u>decrease</u> in the item value.

Here are some examples on how to find these <u>new</u> values:

<u>Example 1</u>:   Jackets are on sale.

They are 25% off the original £80.

What is the <u>discounted</u> price?

<u>Answer</u>:   £80 is 100%.

To find 25%, we do £80 ÷ 4 = **£20**.

The discounted price for the jackets is:

£80 - **£20** = <u>£60</u>

<u>Example 2</u>:   Jessica earns £20 an hour.

She gets a 10% <u>pay rise.</u>

How much does she earn now each hour?

<u>Answer</u>:   £20 is 100%.

To find 10%, we do £20 ÷ 10 = **£2**.

Her new earnings is:

£20 + **£2** = <u>£22</u>

YOU TRY:

6. How much do the discounted jeans cost?

£60

25% off!

7. The following items are all 50% off.
   Find the discounted prices.

£8

£10

£5

8. Next year the school will have 10% more
   students than this year.

   This year has 200 students.

   How many students will there be next year?

9. The price of cheeseburgers is £1.

The price increases by 40%.

What is the new price?

10.

Abi thinks of the number 20.

She decreases it by 20%.

What is the new number?

11. The farmer expects a 30% increase in the number of strawberries next year.

This year 1000 strawberries were grown.

How many strawberries does he expect next year?

Pre 11+ Maths

Spring Half Term Test

Full Name: ………………… Test Date: ………………..............

Tutor: ……………………… Day of Lessons: ……………………

This test lasts 45 minutes.

Read the questions carefully and try all the questions.
If you cannot do a question, move on to the next one.
If you have finished, check over your answers.

The total marks for this paper is 70.

1) Draw a ring around the even numbers and underline the odd numbers.

57, 93, 14, 27, 201, 300, 19, 480 (8)

2) Add these numbers:

a) 3 9 4 + 2 8 0 **b)** 3 5 2 + 1 9

(2)

3) Find the answers to the following:

a) 223 - 99 **b)** 321 - 123

(2)

4) Find the difference of these numbers:

a) 1000 and 289 **b)** 2014 and 1172

(2)

5)

a) What is the square of 3? (1)

b) Which of these numbers are square numbers? (3)
Put a ring around the right ones.

100, 45, 39, 144, 36, 1

6)   **a)**   What is 2 3 6 × 7 ?   (1)

     b)   What is the product of 7 6 3 and 1 2?   (1)

7)

     a   4 | 7 6   **b**   7 | 9 9 4   **c**   9 | 1 5 0 3

     (3)

8)   Cancel these fractions down to their **SIMPLEST** form   (8)

     a) $\frac{3}{9} =$ **b)** $\frac{2}{20} =$ **c)** $\frac{21}{28} =$ **d)** $\frac{6}{36} =$

     e) $\frac{12}{48} =$ **f)** $\frac{18}{36} =$ **g)** $\frac{12}{60} =$ **h)** $\frac{12}{40} =$

9)   Write the following numbers as words: (1)

     a)   2 2 7

     b)   1, 0 5 6   (1)

     c)   7, 4 4 4   (1)

     d)   1 4, 3 6 7   (1)

10) Write the following numbers as figures: (1)

    a) seven hundred and fifty eight

    b) fifteen thousand three hundred and ninety two (1)

    c) eighty three thousand and thirty two (1)

11) Work out the missing numbers: (4)

    a) $\dfrac{9}{16} = \dfrac{}{32}$ **b)** $\dfrac{7}{15} = \dfrac{}{30}$ **c)** $\dfrac{11}{13} = \dfrac{}{39}$ **d)** $\dfrac{1}{7} = \dfrac{}{56}$

12) Find the Lowest Common Multiple (**LCM**) of these pairs of numbers:

    a) 2 and 5   **b)** 3 and 7   **c)** 4 and 9

    (3)

13) Change these fractions to improper. (4)

    a) $1\dfrac{2}{3} =$ **b)** $2\dfrac{3}{7} =$ **c)** $1\dfrac{5}{6} =$ **d)** $1\dfrac{9}{11} =$

14) If Ali has $\dfrac{1}{4}$ of the pocket money I get and he gets £7.00, how much do I get?

    (1)

15) A coat costs £60. (1)
In a sale, its price is reduced by $\frac{1}{3}$

   a) What is $\frac{1}{3}$ of £60?

   b) What price is the coat sold for? (1)

16) Change these fractions to mixed numbers (4)

   a) $\frac{11}{7} =$ **b)** $\frac{19}{5} =$ **c)** $\frac{19}{3} =$ **d)** $\frac{21}{11} =$

17) **a)** Find the Highest Common Factor (HCF) of 16 and 40. (1)

   b) Find the LCM of 17 and 34. (1)

18) **ESTIMATE** the answer to the following problem starting from the left side.

$$275 \div 19 + 745$$

Hint: Start by rounding 275 to the nearest 100.

Then round each number appropriately. (5)

19)   What is $\frac{1}{5}$ of £50?   (1)

20)   What is $\frac{3}{5}$ of £50?   (1)

21)   Write any equivalent fraction to $\frac{3}{8}$ (1)

22)   Kiana ate 8 chocolates in a box which left me with $\frac{3}{4}$ of the box.
   a)   How many in the box to start with?   (1)

   b)   How many did I get?   (1)

23)   Which is more,
   a) $\frac{1}{2}$ of 64 kg,

   b)   or $\frac{5}{8}$ of 64kg?   (2)

Section 2
Topic 9 : Reading Scales

OBJECTIVES

In this module, you will:

❖ Learn about metric units (grams, litres etc.)

❖ Understand how to read numbers on a scale

❖ Be able to read different types of scales (jugs, thermometers etc.)

KEYWORDS:

Grams	A unit of mass
Kilograms	A unit of mass
Millilitres	A unit of capacity for a liquid
Litres	A unit of capacity for a liquid
°C	A unit of temperature of an object

PREVIOUS KNOWLEDGE:

❖ Addition and Subtraction

❖ Multiplication and Division

Understanding what we mean by "unit"

Roughly speaking, units are abbreviations after numbers.

They tell us what the numbers mean and how we can compare one another.

You will learn more about them next week.

Summary of Metric Units in this Module

To measure:	Units
Mass (solids)	grams (g), kilograms (kg)
Capacity (liquids)	millilitres (ml), litres (l)
Temperature	°C (degrees Celsius)

Reading Scales – Measurements and Mass

You will have probably seen scales before.

However, not all the lines have numbers.

How do we use scales like these?

Let's work through an example on the next page.

Example – Find what value the arrow represents.

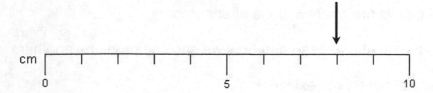

Answer

Step 1

First count how many lines there are as follows:

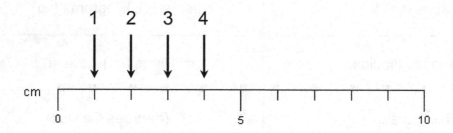

There are 5 lines in between, on the number line, 0 to 5.

Step 2

Find the value for that part of the scale:

$$5 - 0 = \underline{5}$$

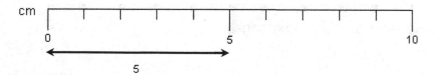

Step 3

Divide the 5 from our scale with the number of lines:

$$5 \div 5 = \underline{1}$$

This means each line on the scale goes up by <u>1</u> each time.

Step 4

Labelling the numbers:

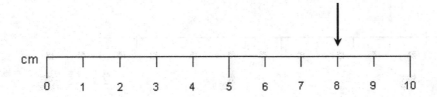

The value that the arrow represents is <u>8</u>.

The units of the scale are on the left: "<u>cm</u>".

With the units, the arrow represents <u>8cm</u>.

YOU TRY:

1. For the following, find the value of the arrow on the scale. Remember to write your answers into your books.

a)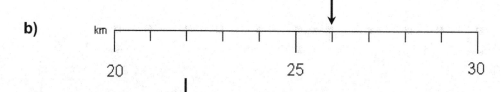

b)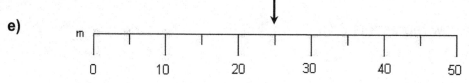

c)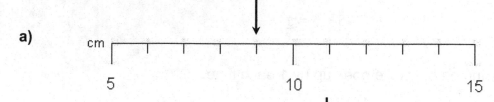

d)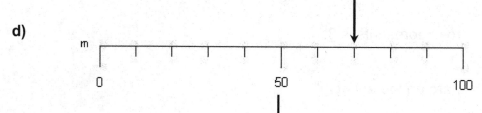

e)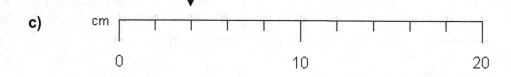

f)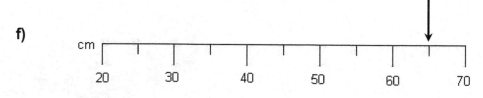

g) h)

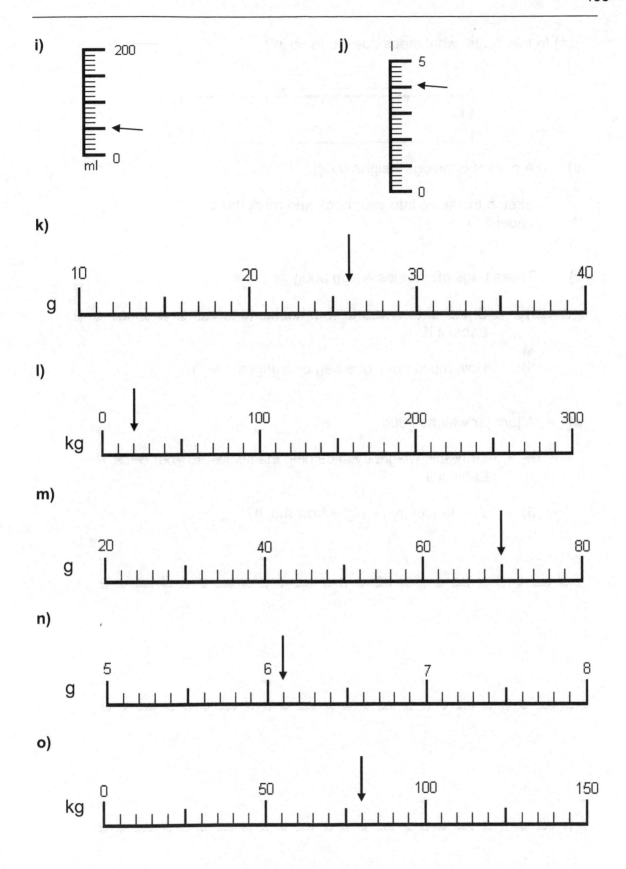

i)

j)

k)

l)

m)

n)

o)

2 a) In this scale, what steps does it go up in?

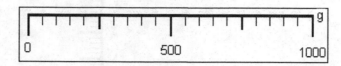

b)   A packet of onions weighs 800g.

Sketch the scale into your book and mark this weight.
Label it O.

c)   Three bags of oranges weigh 600g.

i)   Sketch the scale into your book and mark this weight.
Label it K

ii)   How much does one bag of oranges weigh?

d)   A jam jar weighs 200g.

i)   Mark this weight by sketching the scale into your book.
Label it J.

ii)   Two jars of jam weighs how much?

Reading Scales – Capacity

Example – How much water does the jug have in it?

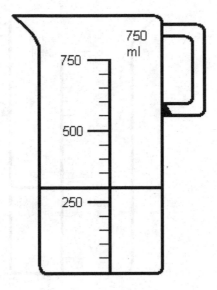

Step 1

First count how many lines there are as follows:

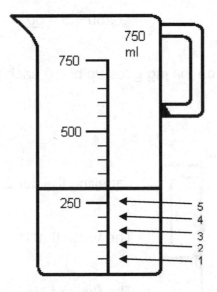

There are 5 lines in between, on the jug, the bottom (0ml) to 250ml.

Step 2

Find the value for that part of the scale:

$$250 - 0 = \underline{250}$$

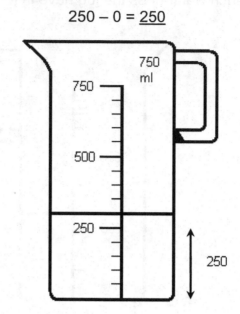

Step 3

Divide the 250 from our jug with the number of lines:

$$250 \div 5 = \underline{50}$$

This means each line on the jug goes up by <u>50</u> each time.

Step 4

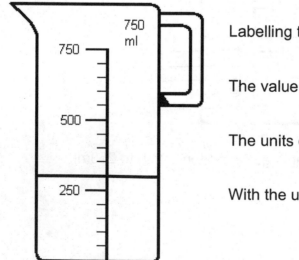

Labelling the numbers:

The value that the arrow represents is <u>300</u>.

The units of the jug are <u>ml</u>.

With the units, the arrow represents <u>300ml</u>.

YOU TRY

5.   How much water is there in each jug?

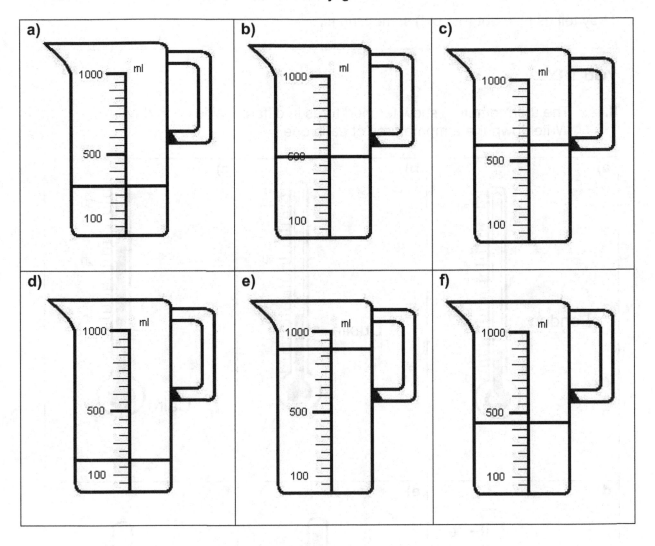

6.   Which one of the jugs is <u>half</u> full?

7.   How much <u>more</u> water would you have to fill each jug to make up 1000ml?
      Write the amount for each of the questions a) to f).

Reading Thermometers

<u>Thermometers</u> are scales that measure temperature.

They tell us how hot (or cold) something is.

YOU TRY

8.   The thermometers show temperatures in different cities one day.
       Write down the temperatures of each one.

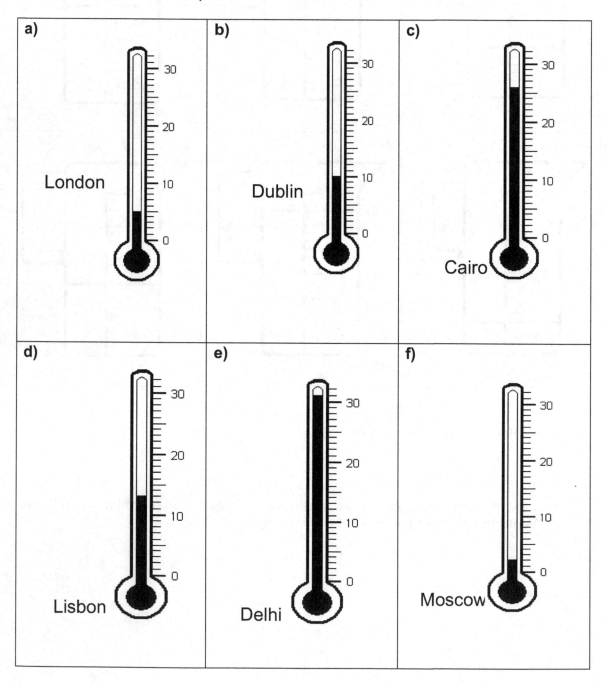

a) London

b) Dublin

c) Cairo

d) Lisbon

e) Delhi

f) Moscow

9. <u>Copy and shade</u> these thermometers to show the correct temperatures. Use a ruler to help you draw the scales.

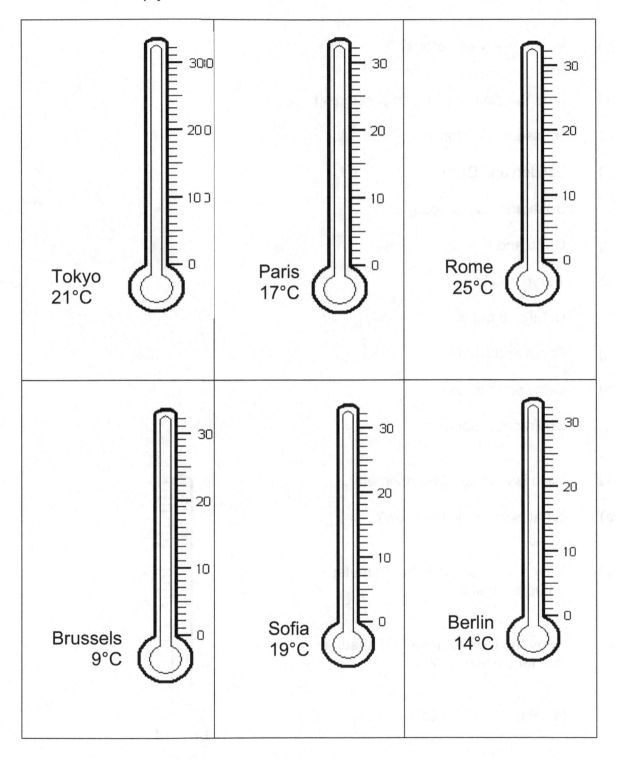

Tokyo
21°C

Paris
17°C

Rome
25°C

Brussels
9°C

Sofia
19°C

Berlin
14°C

10. Look at all twelve cities.

a) Which one was coldest?

b) Which one was hottest?

11. Find the difference in temperatures between:

a) Brussels and Moscow

b) London and Delhi

c) Sofia and Luxembourg

d) Cairo and Rome

e) Tokyo and Lisbon

f) Delhi and Sofia

g) Berlin and Dublin

h) Cairo and London

i) Berlin and Lisbon

12. Here is a different thermometer.

a) What is value of the arrow?

b) How much does this thermometer go up each step?

c) Copy the thermometer and shade in the temperature 52°C.

d) What is 52°C - 22°C?

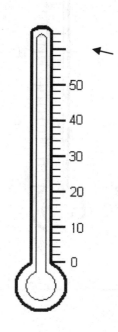

Section 2
Topic 10 : Units - Length

OBJECTIVES

In this module, you will:

❖ Understand about lengths and measuring with a ruler

❖ Learn about the units millimetres (mm), centimetres (cm), meters (m) and

kilometres (km)

❖ Understand how to convert between mm and cm

❖ Understand how to convert between cm and m

❖ Understand how to convert between m and km

KEYWORDS:

Millimetres (mm)	A unit of length (10 mm = 1 cm)
Centimetres (cm)	A unit of length (100 cm = 1 m)
Meters (m)	A unit of length (1000 m = 1 km)
Kilometres (km)	A unit of length

PREVIOUS KNOWLEDGE:

❖ Addition and Subtraction
❖ Multiplication and Division
❖ Reading Scales

Recap on the word "unit"

Roughly speaking, units are abbreviations after numbers.

They tell us what the numbers mean and how we can compare one another.

Measuring lengths

Lengths tell us how long or short something is.

We can measure lengths using a ruler:

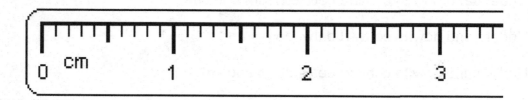

Each line represents 0.1cm (centimetres).
If you don't understand, check back on Reading Scales last week.

Example: Measure how long this line is in cm.

Answer: Place your ruler so that it is in line with 0:

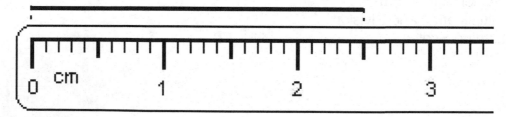

According to this ruler, the line is 2.5cm long.

YOU TRY:

1.   How long are these lines?
       Measure them with your ruler.
       Write your answers in centimetres (cm) into your books clearly.

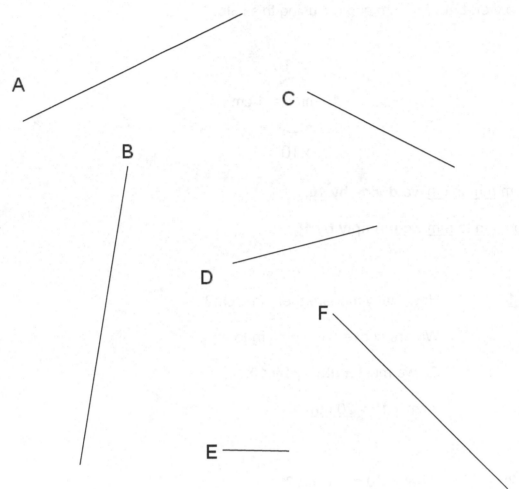

2.   Draw lines for these lengths:

       2.5 cm

       5.0 cm

       1.2 cm

       3.4 cm

Millimetres and centimetres

We know <u>millimetres (mm)</u> and <u>centimetres (cm)</u> are both <u>units of lengths</u>. They are used for <u>small</u> lengths such as books and pens.

There are 10 millimetres in 1 centimetre.

We can convert between mm and cm using this **rule**:

$$\div 10$$

$$10 \text{ mm} = 1 \text{ cm}$$

$$\times 10$$

To get from <u>mm</u> to <u>cm</u> we **divide by 10**.

To get from <u>cm</u> to <u>mm</u> we **multiply by 10**.

<u>Example 1:</u>   How many mm are there in 2 cm?

<u>Answer:</u>   We are converting from <u>cm</u> to <u>mm</u>.

So we must **multiply by 10**.

2 cm x 10 = <u>**20 mm**</u>

<u>Example 2:</u>   What is 30 mm in cm?

<u>Answer:</u>   We are converting from <u>mm</u> to <u>cm</u>.

So we must **divide by 10**.

30 mm ÷ 10 = <u>**3 cm**</u>

<u>Note:</u> It is **<u>very important</u>** to write down the **<u>units</u>** with your answer

3.   Change these mm into cm.

a)	10 mm	**i)**	25 mm
b)	70 mm	**j)**	32 mm
c)	60 mm	**k)**	16 mm
d)	30 mm	**l)**	28 mm
e)	90 mm	**m)**	85 mm
f)	50 mm	**n)**	5 mm
g)	20 mm	**o)**	8 mm
h)	15 mm	**p)**	2 mm

4.   Change these cm into mm.

a)	4 cm	**g)**	3.6 cm
b)	6 cm	**h)**	8.1 cm
c)	7 cm	**i)**	23 cm
d)	9 cm	**j)**	34 cm
e)	1.5 cm	**k)**	17 cm
f)	2.5 cm	**l)**	68 cm

5.   This ruler goes up by 1mm each time.

What values do the arrows point to?

Copy the sentences and fill in the blank spaces.

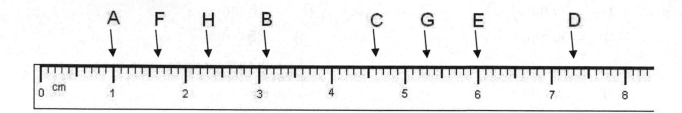

A points to <u>1</u> cm, which is the same as <u>10</u> mm.

B points to <u>3.1</u> cm, which is the same as __ mm.

C points to <u>4.6</u> cm, which is the same as __ mm.

D points to __ cm, which is the same as __ mm.

E points to __ cm, which is the same as __ mm.

F points to __ cm, which is the same as __ mm.

G points to __ cm, which is the same as __ mm.

H points to __ cm, which is the same as __ mm.

Centimetres and Metres

Similarly <u>metres (m)</u> is also a <u>unit of length</u>.

They are used for <u>quite long</u> lengths such as the length of your room or your bed.

There are 100 centimetres in 1 metre.

We can convert between cm and m using a **similar rule**:

$$\div 100$$

100 cm = 1 m

$$\times 100$$

To get from <u>cm</u> to <u>m</u> we **divide by 100**.

To get from <u>m</u> to <u>cm</u> we **multiply by 100**.

<u>Example 1:</u>	How many cm are there in 3 m?
<u>Answer:</u>	We are converting from <u>m</u> to <u>cm</u>.
	So we must **multiply by 100**.
	3 m x 100 = <u>**300 cm**</u>
<u>Example 2:</u>	What is 200cm in m?
<u>Answer:</u>	We are converting from <u>cm</u> to <u>m</u>.
	So we must **divide by 100**.
	200 cm ÷ 100 = <u>2 m</u>
<u>Reminder:</u>	It is **important** to write down the <u>units</u> with your answer

YOU TRY:

6. Change these m into cm.

a)	1 m		**g)**	1.2 m
b)	2 m		**h)**	1.9 m
c)	5 m		**i)**	6.2 m
d)	9 m		**j)**	9.1 m
e)	2.5 m		**k)**	4.6 m
f)	3.0 m		**l)**	10.7 m

7. Convert these measurements in cm into m.

a)	100 cm		**g)**	410 cm
b)	200 cm		**h)**	624 cm
c)	800 cm		**i)**	378 cm
d)	500 cm		**j)**	90 cm
e)	250 cm		**k)**	55 cm
f)	330 cm		**l)**	12 cm

The following questions show how you can split up your measurements into tens and hundreds to see how a <u>metre</u> is made up.

8. Use the example to change the measurements in m into cm.
Copy and complete the sentences.

a) 1 m 25 cm = <u>100</u> + <u>25</u> = <u>125</u> cm

b) 3 m 10 cm = ___ + ___ = ____ cm

c) 2 m 30 cm = ___ + ___ = ____ cm

d) 7 m 55 cm = ___ + ___ = ____ cm

e) 6 m 14 cm = ___ + ___ = ____ cm

9.   Copy and complete.

a)   Haris is 1 metre and 63 centimetres tall.
What is his height in centimetres?

1 m 63 cm = ___ + ___ = ____ cm

b)   Humara is 1 metre and 52 centimetres tall.
What is her height in centimetres?

1 m 52 cm = ___ + ___ = ____ cm

10.   Copy and complete the sentences.
Use the example to change these cm into m.

a)   175 cm = <u>100</u> + <u>75</u> = <u>1</u> m <u>75</u> cm = <u>1.75</u> m

b)   220 cm = ___ + ___ = __ m __ cm = ____ m

c)   352 cm = ___ + ___ = __ m __ cm = ____ m

d)   425 cm = ___ + ___ = __ m __ cm = ____ m

e)   703 cm = ___ + ___ = __ m __ cm = ____ m

f)   198 cm = ___ + ___ = __ m __ cm = ____ m

Metres and Kilometres

Similarly kilometres (km) is also a unit of length.

They are used for long lengths such as a race track.

There are 1000 metres in 1 kilometre.

We can convert between m and km using a **similar rule**:

$$\div 1000$$
$$1000 \text{ m} = 1 \text{ km}$$
$$\times 1000$$

To get from m to km we **divide by 1000**.

To get from km to m we **multiply by 1000**.

Example 1:   How many m are there in 4 km?

Answer: We are converting from km to m.

                So we must **multiply by 1000**.

                4 km x 1000 = <u>**4000 m**</u>

Example 2:   What is 2000m in km?

Answer: We are converting from m to km.

                So we must **divide by 1000**.

                2000 m ÷ 1000 = <u>**2 km**</u>

Reminder: It is **important** to write down the units with your answer

YOU TRY:

11. Change these km into m.

a) 1 km

b) 3

c) 7 km

d) 2 km

e) 5.5 km

f) 4.0 km

g) 1.6 km

h) 8.3 km

i) 7.2 km

j) 5.35 km

k) 7.13 km

l) 9.62 km

12. Convert these measurements in m into km.

a) 1000 m

b) 4000 m

c) 5000 m

d) 7000 m

e) 6500 m

f) 3300 m

g) 4150 m

h) 3650 m

i) 3700 m

j) 400 m

k) 350 m

l) 10 m

13. This scale shows one kilometre split into ten parts.
If the question is in km, write it in m.
If the question is in m, write it in km.

a) 0.2 km

b) 0.8 km

c) 0.7 km

d) 200 m

e) 600 m

f) 0.1 km

g) 1.0 km

h) 0.6 km

i) 100 m

j) 2.7 km

k) 9.5 km

l) 3750 m

m) 2500 m

km 1.0 ——— 1000 m
0.9 ——— 900
0.8 ——— 800
0.7 ——— 700
0.6 ——— 600
0.5 ——— 500
0.4 ——— 400
0.3 ——— 300
0.2 ——— 200
0.1 ——— 100
0 ——— 0

Adding Lengths

Now that you know how to convert between <u>units</u>, we can add lengths with <u>different</u> units together.

<u>Example</u>: Add 22 mm and 1 cm.

<u>Answer</u>: We know 22 mm is the same as <u>2.2 cm</u>.

22 mm + 1 cm

= 2.2 cm + 1 cm

= <u>3.2 cm</u>

<u>Important</u>: You <u>must</u> have the <u>same</u> units <u>before</u> you add.

YOU TRY:

14. Add the measurements together.
Copy and fill in the spaces.

a) 10 mm + 2 cm = ___ cm + 2 cm = _____ cm

b) 15 mm + 4 cm = ___ cm + 4 cm = _____ cm

c) 25 mm + 5 cm = 25 mm + ___ mm = _____ mm

d) 100 cm + 3 m = ___ m + 3 m = _____ m

e) 300 cm + 2 m = ___ m + 2 m = _____ m

f) 450 cm + 5 m = 450 cm + ___ cm = _____ cm

g) 1000 m + 1 km = ___ km + 1 km = _____ km

h) 2000 m + 3 km = ___ km + 3 km = _____ km

i) 3500 m + 5 km = 3500 m + _____ m = _____ m

Section 2
Topic 11 : Perimeter

OBJECTIVES

In this module, you will:

❖   Understand what we mean by **perimeter**

❖   Be able to find the perimeters of **regular** shapes

❖   Be able to find the perimeters of **irregular** shapes

KEYWORDS:

Perimeter	The total distance around the outside of a shape
Regular Shape	A shape that has both equal angles and equal sides
Irregular Shape	A shape that does not have both equal angles and equal sides.

PREVIOUS KNOWLEDGE:

❖   Addition and Subtraction
❖   Multiplication and Division
❖   Units - Lengths

Perimeters of Regular Shapes

The **perimeter** of a shape is the total distance around the outside of a shape.

A **regular shape** has equal angles and equal sides.

Here are some regular shapes:

Triangle   Square   Pentagon   Hexagon

Regular shapes have equal sides.

This means all the sides have the same length.

Example: Find the perimeter of this
                regular triangle.

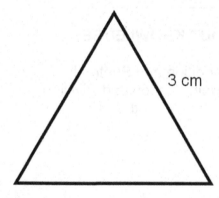

3 cm

Answer: We know a triangle has 3 sides.

                It is also a regular triangle. This means all 3 sides have the
                same length of 3cm.

                Perimeter: 3 cm x 3 = 9 cm

YOU TRY:

1. Find the <u>perimeters</u> of these <u>regular</u> shapes.

a)

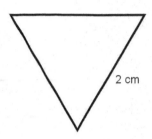

Perimeter: _____ cm

d)

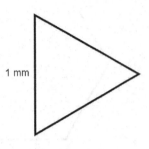

1 mm

Perimeter: _____ mm

b)

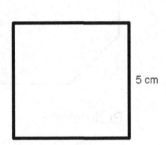

5 cm

Perimeter: _____ cm

e)

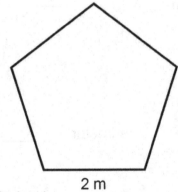

2 m

Perimeter: _____ m

c)

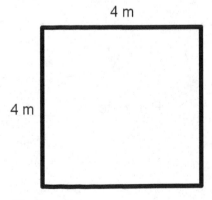

4 m

4 m

Perimeter: _____ m

f)

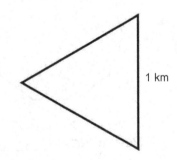

1 km

Perimeter: _____ km

2. Count the number of sides carefully and find the perimeter of each shape. Write down the perimeter.

a)

2 cm

Perimeter: _____ cm

b)

1 mm

Perimeter: _____ mm

c)

4 km

Perimeter: _____ km

d)

3 km

Perimeter: _____ km

e)

3 m

Perimeter: _____ m

f)

7 cm

Perimeter: _____ cm

Perimeters of Irregular Shapes

An **irregular shape** is one that does not have equal sides or equal angles.

Let's work through an example to see how to find the perimeter of these kinds of shapes.

Example 1:   Find the perimeter of this shape.

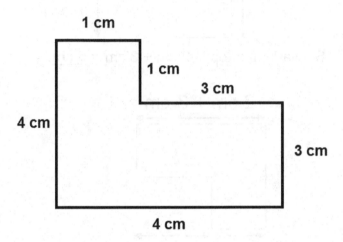

Answer: To find the perimeter, we must add all the lengths of the sides together:

1 cm + 1 cm + 3 cm + 3 cm + 4 cm + 4 cm

=   16 cm

Note: You can add the numbers in any order.

Example 2:

a)   Find out the missing lengths for this irregular shape.

b)   Find the perimeter of the shape.

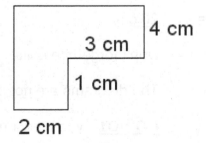

Answer:

a)   Let's look at the left side first.

The lengths on the right show 4 cm and 1 cm.

The left side is the <u>sum</u> of 4 cm and 1 cm, which is <u>5 cm</u>.

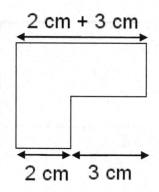

Similarly, the top is 2 cm + 3 cm = <u>5 cm</u>.

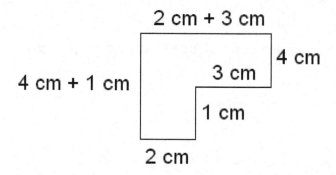

b)   We can now work out the perimeter:

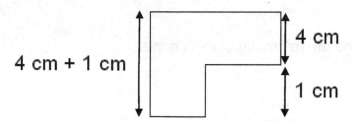

<u>Perimeter</u>:   (2 cm + 3 cm) + 4 cm + 3 cm + 1 cm + 2 cm + (4 cm + 1 cm)

=   5 cm   + 4 cm + 3 cm + 1 cm + 2 cm + 5 cm

=   <u>20 cm</u>

<u>Note 1</u>:   Don't forget the units!

<u>Note 2</u>:   The diagrams are <u>not</u> drawn to scale.

<u>DO NOT</u> try to measure them with your ruler!

YOU TRY:

3. Find the perimeter of these shapes.

a)

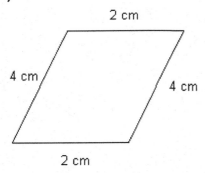

2 cm

4 cm

4 cm

2 cm

e)

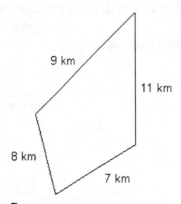

9 km

11 km

8 km

7 km

b)

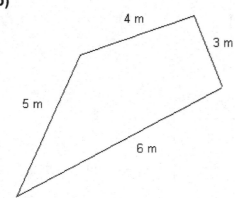

4 m

3 m

5 m

6 m

f)

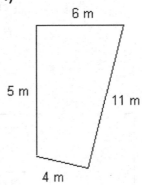

6 m

5 m

11 m

4 m

c)

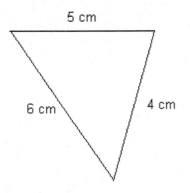

5 cm

6 cm

4 cm

g)

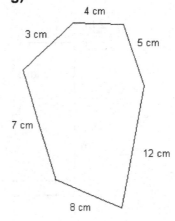

4 cm

3 cm

5 cm

7 cm

12 cm

8 cm

d)

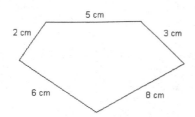

5 cm

2 cm

3 cm

6 cm

8 cm

h)

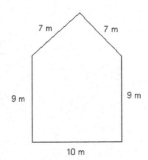

7 m

7 m

9 m

9 m

10 m

4. For each question:

- <u>Sketch</u> the diagram.

- Find <u>all</u> the missing lengths on the shape
- (<u>**Do not**</u> **measure – you will get it wrong**)

- Find the perimeter.

a)

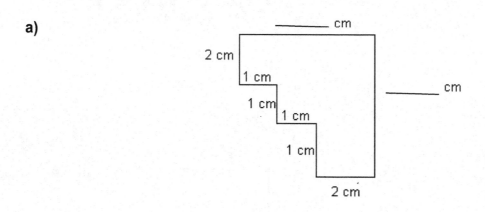

b)

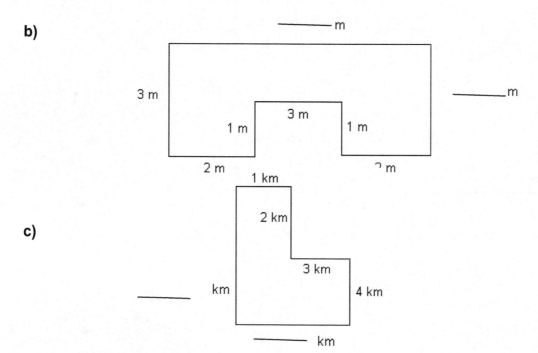

c)

d)

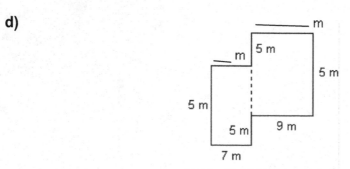

e)

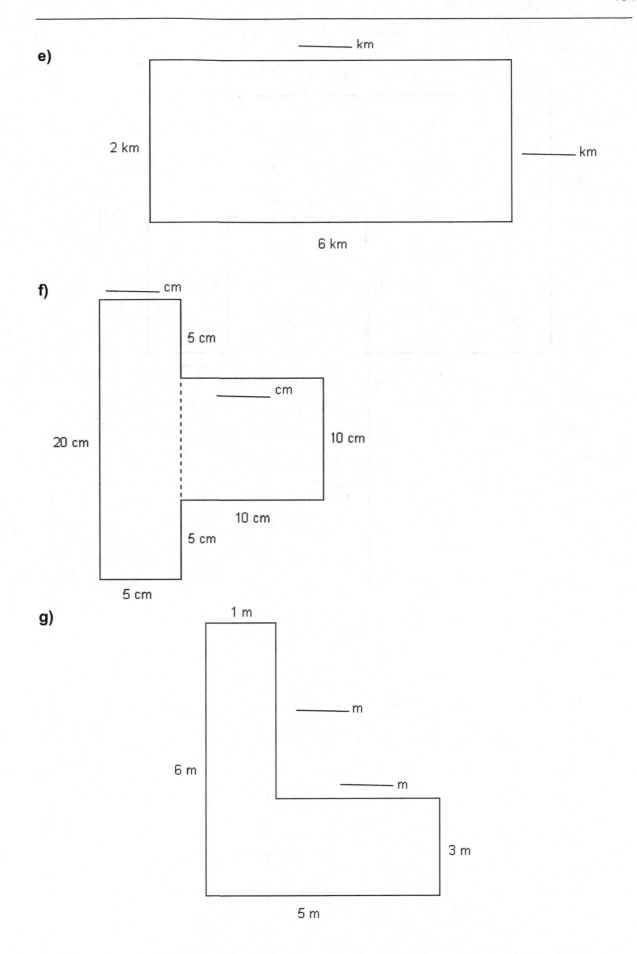

_____ km

2 km

6 km

_____ km

f)

_____ cm

5 cm

_____ cm

20 cm

10 cm

10 cm

5 cm

5 cm

g)

1 m

_____ m

6 m

_____ m

3 m

5 m

h)

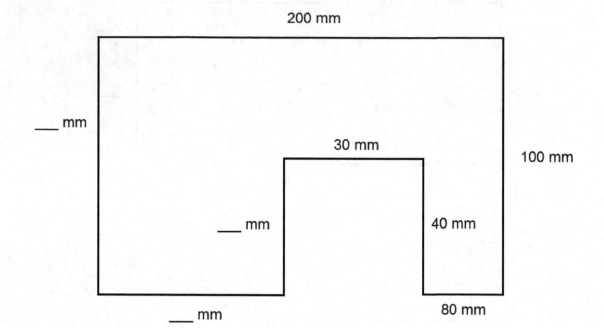

200 mm

___ mm

30 mm

100 mm

___ mm

40 mm

___ mm

80 mm

Section 2
Topic 12: Units - Mass and Volume

OBJECTIVES

In this module, you will:

❖   Understand what is mass and volume

❖   Learn about the units grams (g), kilograms (kg), millilitres (ml) and litres (ℓ)

❖   Understand how to convert between g and kg

❖   Understand how to convert between ml and ℓ

KEYWORDS:

Grams (g)   A unit of mass (1000 g = 1 kg)

Kilograms (kg)   A unit of mass

Millilitres (ml)   A unit of volume (1000 ml = 1 ℓ)

Litres (ℓ)   A unit of volume

PREVIOUS KNOWLEDGE:

❖   Reading Scales
❖   Units – Lengths
❖   Multiplication and Division

Grams and Kilograms

Grams (g) and kilograms (kg) are <u>units of mass</u>.

Mass tells us how much "stuff" there is in an object.

We use grams for <u>lighter</u> objects such as fruit and vegetables.

We use kilograms for <u>heavier</u> objects such as cars and people.

There are 1000 grams in 1 kilogram.

Similar to last week, we can convert between g and kg using this **rule**:

$$\div 1000$$
$$1000 \text{ g} = 1 \text{ kg}$$
$$\times 1000$$

To get from <u>g</u> to <u>kg</u> we **divide by <u>1000</u>**.

To get from <u>kg</u> to <u>g</u> we **multiply by <u>1000</u>**.

<u>Example 1:</u>   How many g are there in 1.9 kg?

<u>Answer:</u>   We are converting from <u>kg</u> to <u>g</u>.

       So we must **multiply by 1000**.

       1.9 kg x 1000 = <u>**1900 g**</u>

<u>Example 2:</u>   What is 5250 g in kg?

<u>Answer:</u>   We are converting from <u>g</u> to <u>kg</u>.

       So we must **divide by 1000**.

       5250 g ÷ 1000 = <u>**5.25 kg**</u>

YOU TRY:

1. Change these kg into g.

a) 1 kg

b) 2 kg

c) 6 kg

d) 7 kg

e) 3.5 kg

f) 6.5 kg

g) 3.2 kg

h) 4.8 kg

i) 2.32 kg

j) 5.79 kg

k) 1.43 kg

l) 3.26 kg

2. Convert these masses in g into kg.

a) 1000 g

b) 5000 g

c) 6000 g

d) 3000 g

e) 2500 g

f) 1500 g

g) 1250 g

h) 3500 g

i) 6700 g

j) 1376 g

k) 150 g

l) 50 g

3. Do you think these weights are sensible? Write <u>Yes</u> or <u>No</u> into your books.

a) An average person weighs 60 g.

b) This booklet of paper weighs 80 g.

c) A lorry weighs around 200 kg.

d) The weight of a feather is 4 g.

e) A small packet of strawberries weighs 5 kg.

4.

The following table shows the cost to post parcels.

Weight not over	Price
1 kg	£3.00
1.5 kg	£5.00
2 kg	£6.50
4 kg	£7.50
6 kg	£8.00

Copy the sentences and write down the price for posting parcels with these weights.

The first one has been done for you.

a) A 3 kg parcel costs <u>£7.50</u> as it weighs less than <u>4 kg</u>.

b) A 5 kg parcel costs £____ as it weights less than ___ kg.

c) A 1.2 kg parcel costs £____ as it weights less than ___ kg.

d) A 1.75 kg parcel costs £____ as it weights less than ___ kg.

e) A 2.1 kg parcel costs £____ as it weights less than ___ kg.

5. What units would you use to weigh these items?
Choose between <u>g</u> and <u>kg</u>.

a) A dog

b) A coin

c) A parcel of books

d) A packet of grapes

e) An elephant

6. This scale measures up to 8 kg.
It goes up in 0.1 kg intervals.

Copy the sentences and write down the values in both kg and g for the following arrows.

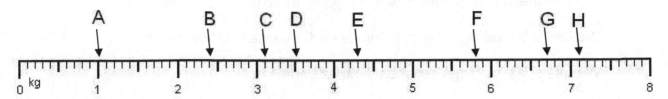

A points to <u>1</u> kg, which is the same as <u>1000</u> g.

B points to <u>2.4</u> kg, which is the same as _____ g.

C points to <u>3.1</u> kg, which is the same as _____ g.

D points to _____ kg, which is the same as _____ g.

E points to _____ kg, which is the same as _____ g.

F points to _____ kg, which is the same as _____ g.

G points to _____ kg, which is the same as _____ g.

H points to _____ kg, which is the same as _____ g.

7. By following the example, copy and complete the sentences filling in the spaces.

a) 1.5 kg is <u>1</u> kg <u>500</u> g, or <u>1500</u> g.

b) 3.4 kg is __ kg _____ g, or _____ g.

c) 6.2 kg is __ kg _____ g, or _____ g.

d) 5.1 kg is __ kg _____ g, or _____ g.

e) _____ kg is __ kg _____ g, or <u>3200</u> g.

f) _____ kg is __ kg _____ g, or <u>6500</u> g.

g) _____ kg is __ kg _____ g, or <u>1300</u> g.

h) _____ kg is <u>2</u> kg <u>500</u> g, or _____ g.

i) _____ kg is <u>8</u> kg <u>200</u> g, or _____ g.

j) _____ kg is <u>7</u> kg <u>300</u> g, or _____ g.

Millilitres and Litres

Millilitres (ml) and litres (ℓ) are <u>units of volume</u>.

We use these to measure how much <u>liquid</u> there is in a container.

We use millimeters for <u>small</u> volumes such as small bottles of water.

We use litres for <u>larger</u> volumes such as barrels of oil.

There are 1000 millilitres in 1 litre.

We can convert between ml and I using this **rule:**

$$\div 1000$$

$$1000 \text{ ml} = 1 \, \ell$$

x1000

To get from <u>ml</u> to <u>ℓ</u> we **divide by <u>1000</u>.**

To get from <u>ℓ</u> to <u>ml</u> we **multiply by <u>1000</u>.**

<u>Example 1</u>:	How many ml are there in 3.1 ℓ?
<u>Answer</u>:	We are converting from <u>ℓ</u> to <u>ml</u>.
	So we must **multiply by 1000**.
	3.1 ℓ x 1000 = <u>**3100 ml**</u>
<u>Example 2</u>:	What is 4500 ml in ℓ?
<u>Answer</u>:	We are converting from <u>ml</u> to <u>ℓ</u>.
	So we must **divide by 1000.**
	4500 ml ÷ 1000 = <u>4.5 ℓ</u>

YOU TRY:

8. Change these ℓ into ml.

a) 1 ℓ

b) 3 ℓ

c) 6 ℓ

d) 5 ℓ

e) 1.5 ℓ

f) 8.5 ℓ

g) 6.1 ℓ

h) 3.3 ℓ

i) 7.98 ℓ

j) 6.43 ℓ

k) 2.51 ℓ

l) 8.02 ℓ

9. Convert these volumes in ml into ℓ.

a) 1000 ml

b) 4000 ml

c) 2000 ml

d) 9000 ml

e) 3500 ml

f) 4500 ml

g) 1250 ml

h) 2500 ml

i) 7100 ml

j) 1176 ml

k) 250 ml

l) 30 ml

10. Order these sets of volumes starting with the largest.

<u>Convert</u> them first if you need to.

a)　　1000 ml　　　1250　ml　　　990 ml　　　1200 ml　　　1005 ml

b)　　10 ml　　　17 ml　　　6.3 ℓ　　　150 ml　　　2.2 ℓ

c)　　200 ml　　　1.2 ℓ　　　3.1 ℓ　　　1.9 ℓ　　　3200 ml

d)　　1.5 ℓ　　　1700 ml　　　1650 ml　　　1.9 ℓ　　　2.6 ℓ

11. Estimate how many litres these items hold:

a)　　A large bottle of water

b)　　A can of coke

c)　　A small bottle of Fanta

d)　　A barrel of wine

e)　　A car's petrol tank

12. **Copy** and **complete** the sentences to fill in the spaces.

Here is a peculiar scale that starts to increase downwards.

The scale goes up in _____ ℓ, which is the same as 100 ml.

a) **A** points to 0.4 ℓ, which is the same as 400 ml.

b) **B** points to ___ ℓ, which is the same as _____ ml.

c) **C** points to ___ ℓ, which is the same as _____ ml.

d) **D** points to ___ ℓ, which is the same as _____ ml.

e) **E** points to ___ ℓ, which is the same as _____ ml.

Copy the scale and draw arrows to show where these volumes are on the scale.

f) **F** points to 1.1 ℓ.

g) **G** points to 3.5 ℓ.

h) **H** points to 7000 ml.

i) **I** points to 6.6 ℓ.

j) **J** points to 7900 ml.

k) **K** points to 4100 ml.

Pre 11+ Maths

Spring End of Term Test

Full Name: …………………… Test Date: …………………................

Tutor: ……………………………… Day of Lessons: …………….…………

This test lasts 30 minutes.

Read the questions carefully and try all the questions.
If you cannot do a question, move on to the next one.
If you have finished, check over your answers.

The total marks for this paper is 34.

1) From the following number sequence,

3   6   9   12   15   18   21

Use one of these words to complete the sentences:

factor, prime, square, multiple, cube

a) Each number is a _____ of 3 (1)

b) 9 is a _____ number (1)

c) Each of 3, 6 and 9 are _____ of 18 (1)

d) 3 is a _____ number (1)

2)

a) Write 1423 to the nearest 10 (1)

b) Write 3659 to the nearest 1000 (1)

c) Write 18 254 to the nearest 1000 (1)

3)

a) Label the **tenths**, **hundredths**, and **thousandths** for the number below. (1)

b) Also, label the **units**, **tens**, and **hundreds**. (1)

152.624

4)

a) Multiply these numbers: 284 and 56. (2)

b) What is 70.62 ÷ 3. (2)

5) Find the percentages in these questions.

a) 10% of 340 (1)

b) 40% of 70 (1)

c) 60% of 80 (1)

d) 20% of 300 (1)

6)

a) $\frac{1}{2}$ of 102   (1)

_____

_____

b) $\frac{2}{5}$ of 35   (1)

_____

_____

7)

a)   Find the HCF of 8 and 12.   (1)

_____

_____

b)   Find the LCM of 6 and 9.   (1)

_____

_____

8)   Put these in order **<u>largest first</u>**:   (2)

      4.32   2.95   4.21   4.12

_____

9)   Calculate the next two numbers in these sequences:

a)   3   6   9   12   ___ ___   (1)

b)   15   22   29   36   ___ ___   (1)

c)   30   27   24   21   ___ ___   (1)

d)   2.1   2.3   2.5   2.7   ___ ___   (1)

10) Measure these two lines.

Write how long they are in <u>both</u> mm and cm.

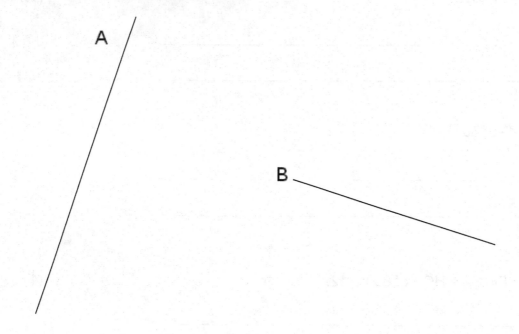

a) Line A is _____ **cm** long, which is the same as _____ **mm**. (2)

b) Line B is _____ **cm** long, which is the same as _____ **mm**. (2)

c) Add the lengths of the lines together.

Write your answer in **mm**. (1)

_____ **mm**

11) Convert into kg:

a) 7391g = _____kg (1)

b) 23g = _____kg (1)

Section 3
Topic 1: Area

OBJECTIVES

In this module, you will:

❖ Understand what we mean by **area**

❖ Be able to find the area of **rectangles**

❖ Be able to find the perimeters of **combined rectangles**

KEYWORDS:

Area	The size of a shape or surface
Rectangle	A shape with opposite sides equal length, four right angles and not all sides the same length
Length	The <u>longer</u> side of the rectangle
Width	The <u>shorter</u> side of the rectangle

PREVIOUS KNOWLEDGE:

❖ Multiplication and Division
❖ Units - Lengths

Area of a Rectangle

The area represents the <u>size</u> of a shape. We will be looking at rectangles:

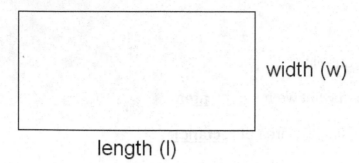

width (w)

length (l)

To find the <u>area of a rectangle</u>, we use this **formula**:

Area (A) = Length (l) x Width (w)

We can use it to find the area of <u>any</u> rectangle.

The <u>length</u> is the <u>longer</u> side.
The <u>width</u> is the <u>shorter</u> side.

<u>**Important:**</u>   The units for area <u>must</u> have ² in it.
             For example, cm² is pronounced "centimetres squared".

<u>Example:</u>   Find the area of this rectangle.

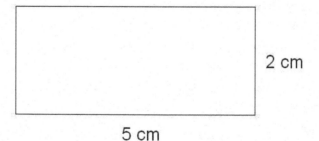

2 cm

5 cm

<u>Answer:</u>   Area = Length x Width   **(Always write the formula)**

             = 5 cm x 2 cm
             = <u>10cm²</u>   **(Note the ² in the units)**

YOU TRY:

1.   Find the area for the following rectangles.

       Don't forget to write the formula at the beginning of each question.

a)

3 cm

8 cm

d)

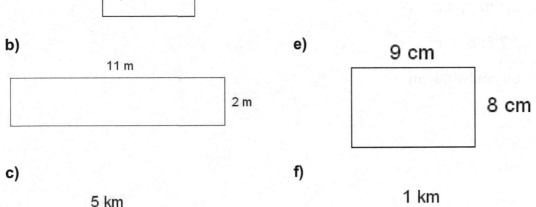

8 m

12 m

b)

11 m

2 m

e)

9 cm

8 cm

c)

5 km

9 km

f)

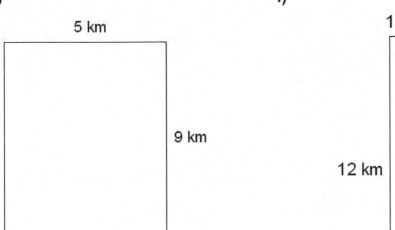

1 km

12 km

2. Find the area of each of these rectangles.
Be careful with your units.

a) 12 cm by 4 cm

b) 9 m by 2 m

c) 8 mm by 7 mm

d) 12 cm by 10 cm

e) 6 m by 11 m

f) 7 mm by 5 mm

g) 5 km by 4 km

h) 2.1 m by 4 m

i) 1.7 m by 5 m

j) 60 cm by 40 cm

3. Look at the following rectangles.

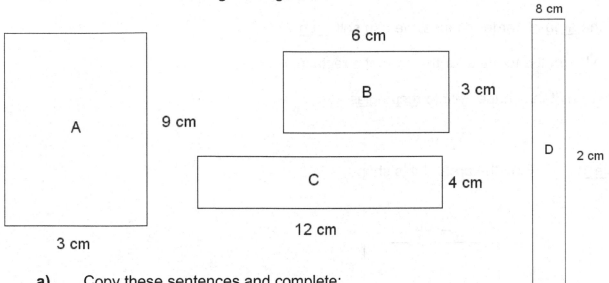

a) Copy these sentences and complete:

Area = length x width.

The area of rectangle A is _____ x _____ = _____ cm²

The area of rectangle B is _____ x _____ = _____ cm²

The area of rectangle C is _____ x _____ = _____ cm²

The area of rectangle D is _____ x _____ = _____ cm²

b) Which rectangle has the greatest area?

c) Which rectangle has the smallest area?

Area of a combined Rectangles

To find the <u>area</u> of harder shapes, we can <u>split it up</u>.

We have learnt the formula for the area of a rectangle.

Let's try to split the shape up into <u>rectangles</u>.

<u>Example 1</u>:   Find the area of this shape.

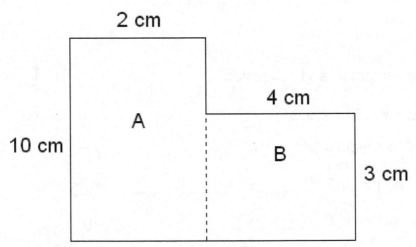

<u>Answer</u>:   Notice this shape is made up of <u>two rectangles</u>: A and B.

<u>Area</u> = <u>length</u> x <u>width</u>.

The area of rectangle A is <u>10 cm</u> x <u>2 cm</u> = <u>20 cm²</u>

The area of rectangle B is <u>4 cm</u>   x <u>3 cm</u> = <u>12 cm²</u>

The total area of the whole shape is the <u>sum of the areas</u> of rectangle A and rectangle B:

<u>20 cm²</u> + <u>12 cm²</u> = <u>32 cm²</u>

Sometimes, you may need to find the <u>missing length</u> before we can find its area.

Example 2:   Find the area of this shape.

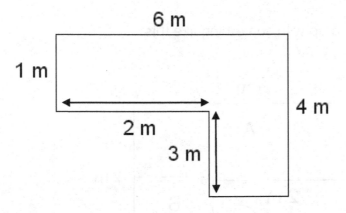

Answer: First let's split the shape up into rectangle A and B:

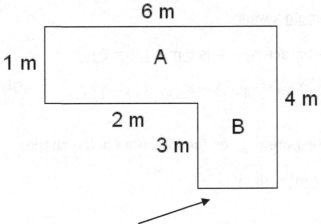

We must find this <u>side</u> before we can find the area of rectangle B.

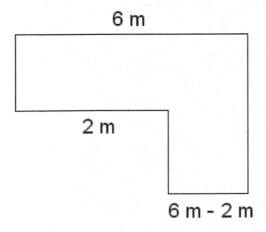

6 m - 2 m

We know the top side is 6 m.

Therefore, the bottom two sides must also add up to 6 m.

The missing side must be 6 m - 2 m = 4 m.

We end up with something like this:

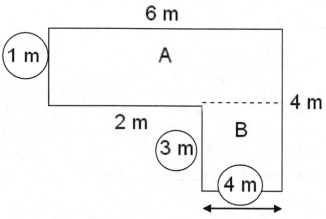

Area = length x width.

The area of rectangle **A** is 6 m x 1 m = 6 m²

The area of rectangle **B** is 4 m x 3 m = 12 m²

Adding their areas gives the total area of the shape:

6 m² + 12 m² = 18 m²

YOU TRY:

4.   For each question, sketch the diagram.

Practice <u>splitting up</u> the shape into <u>two rectangles</u>.
Label them A and B.

Then find the <u>area</u> of the whole shape.

a)

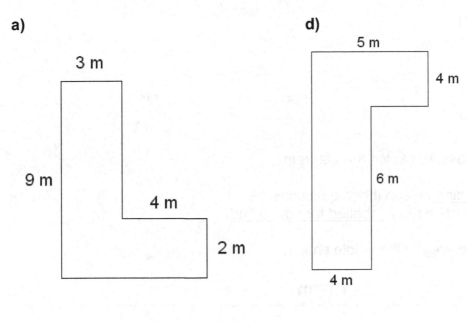

d)

b)

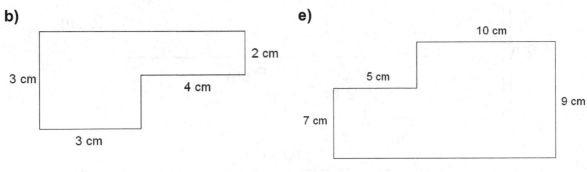

e)

c)

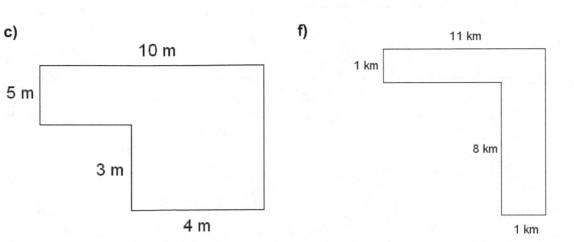

f)

g)

h)

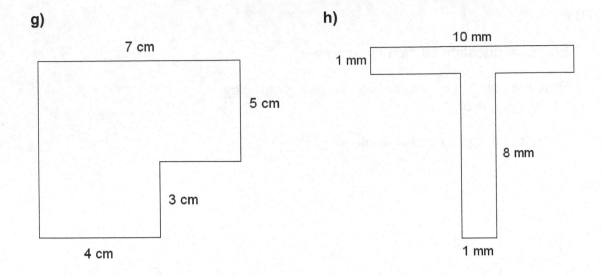

7 cm

5 cm

3 cm

4 cm

10 mm

1 mm

8 mm

1 mm

5.   For each question, sketch the diagram.

Find the <u>missing side</u> in these questions.
The first few have been <u>labelled for you to find</u>.

Then find the <u>area</u> of the whole shape.

a)

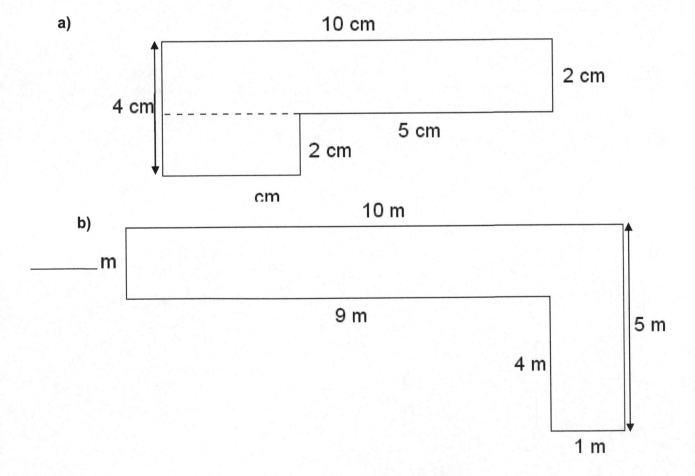

10 cm

2 cm

4 cm

5 cm

2 cm

cm

b)

10 m

m

9 m

5 m

4 m

1 m

c)

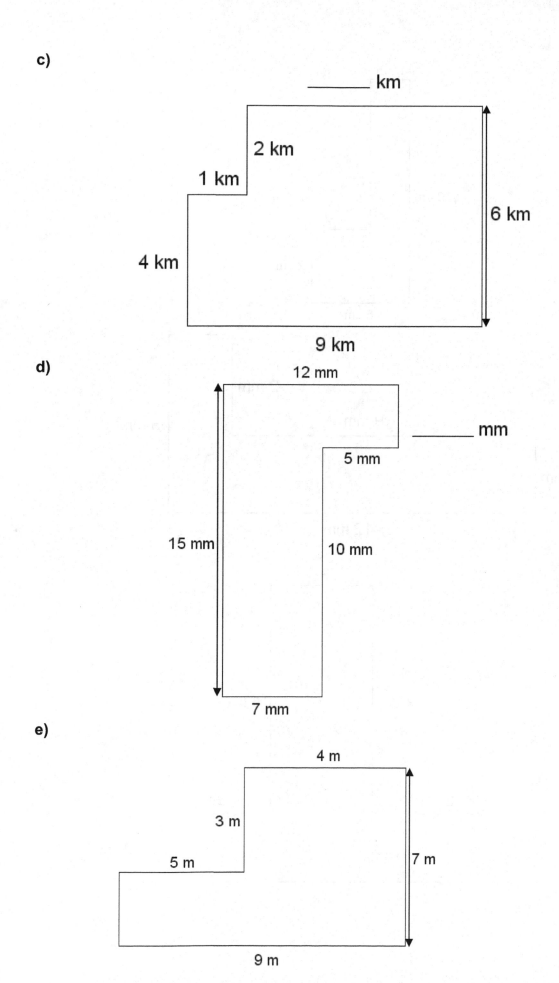

_____ km

2 km

1 km

6 km

4 km

9 km

d)

12 mm

_____ mm

5 mm

15 mm

10 mm

7 mm

e)

4 m

3 m

5 m

7 m

9 m

f)

g)

h)

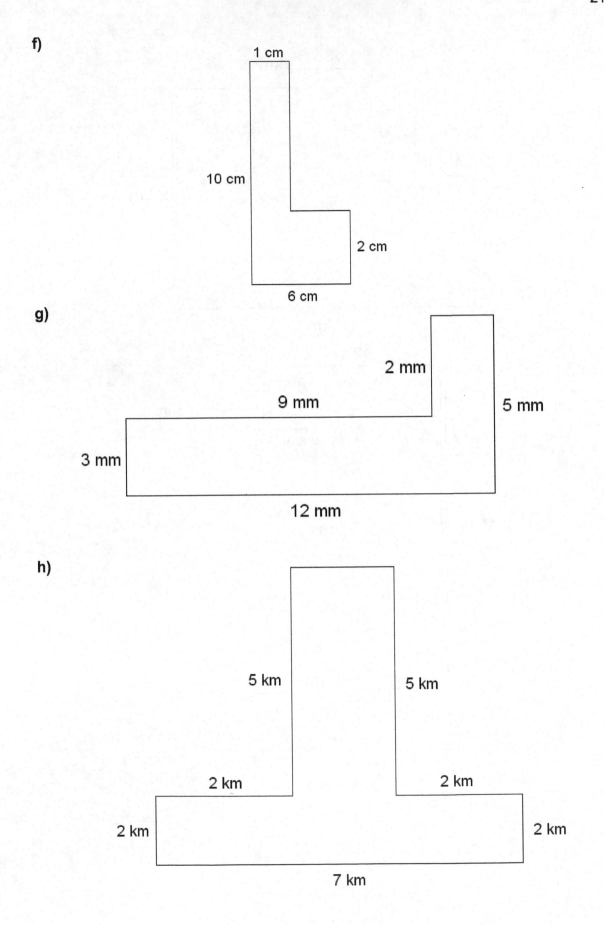

Section 3
Topic 2: Time

OBJECTIVES

In this module, you will:

❖   Understand about clocks

❖   Learn about **analogue** and **digital** forms of time

❖   Be able to write the time in **analogue** format with any number of minutes and hours

❖   Be able to write the time in **digital** format with any number of minutes and hours

❖   Know about and be able to use **am** and **pm** to tell the time of day

---

KEYWORDS:

Hours	Unit of time (hours are the numbers on the clock)
Minutes	Unit of time
Seconds	Unit of time
Analogue Format	The common clock with dials / moving hands
Digital Format	The time shown using numbers only
am	Times that are from midnight to noon
pm	Times that are from noon to midnight

PREVIOUS KNOWLEDGE:

❖   Reading Scales
❖   Addition and Subtraction

Understanding about clocks

● Clocks are **scales** for time.

● There are **three units** of time – **seconds**, **minutes** and **hours**.

● The normal clock uses **three hands** to show them to us:

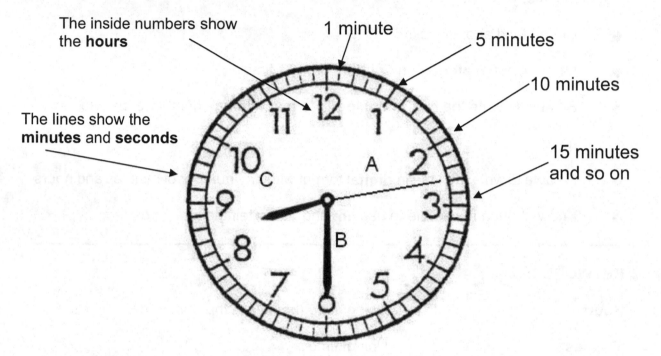

The inside numbers show the **hours**

1 minute

5 minutes

10 minutes

The lines show the **minutes** and **seconds**

15 minutes and so on

● The **short thin hand** (second hand) represents the **seconds**.

● The **long hand** (minute hand) points at the **minutes**.

● The **short hand** (hour hand) tells us the **hour**.

　○ All the hands move **clockwise**.

　○ They move around the clock from 1, to 2, to 3, and so on, all the way to 12 and continues to 1 to repeat.

● There are **60 minutes** in **1 hour**, and there are **60 lines** (have a check!) around the clock.

　○ This means **1 line** is **1 minute**.

　○ The number of lines (and also minutes) goes up in **multiples of 5** to every **numbered hour** on the clock.

Understanding Analogue and Digital Time

Analogue   **Digital**

Analogue: Tells the time using **hands**

Digital: Tells the time using **numbers**

What do you notice about the time the two clocks above show?

_____

Analogue Time – Half and Quarter Hours

We know there are **60 minutes** in **1 hour**.

- **One quarter** of 60 minutes is **15 minutes**.
  - ○ We use this when the **minute hand** points to hours **3** or **9**.

- **One half** of 60 minutes is **30 minutes**.
  - ○ We use this when the **minute hand** points to hour **6**.

Here are some examples:

One o'clock

Quarter past one

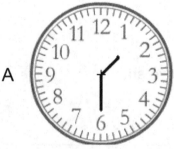

A

Half past one

Quarter to two

Example: For clock labelled A, the minute hand points to the **6**.
This tells us "**half**" of the hour.

The hour hand points in between **one** and **two**.
It is **past one** but not quite reached **two** yet.

Answer: Clock A is **half past one**.

Note: Notice that the last one is **quarter to two**.
Three quarters past one is the **same** as **quarter to two** but normally,
we do not say it that way!

YOU TRY:

1. Write the time for each of the clocks in <u>words</u>.

a)

b)

c)

d)

e)

f)

g)

h)

Analogue Time – Any Minutes

<u>Reminder</u>:

- There are **60 minutes** in **1 hour**
- There are **60 lines** around the clock
- **One line** on the clock tells us **one minute**

We can therefore find **any number of minutes**:

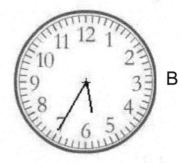

B

Ten past three

Twenty five to six

Twenty five past nine

Three to eleven

<u>Example</u>: For clock B, the minute hand points to the **7**.

Because the minutes go up in multiples of **5** to every numbered hour on the clock, we calculate:

5 x 7 = 35 minutes out of the whole **60** minutes.

It points past the 6 (half way)
We must **subtract** to find the number of minutes "**to**" the **6**th **hour** (shown by the hour hand):

60 – 35 = 25 minutes.

<u>Answer</u>: Clock B is **twenty five to six**

YOU TRY:

2.   Copy the sentences and fill in the blank spaces.
Draw a clock to show the time on the right hand side.

a)

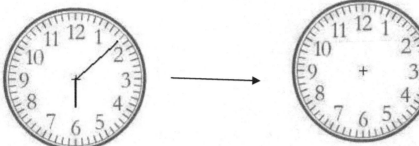

The time is…

Draw a clock to show:
In half an hour, it will be…

b)

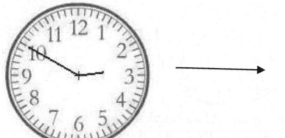

The time is…

Draw a clock to show:
In twenty minutes, it will be…

c)

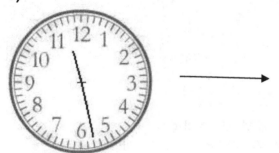

Draw a clock to show:
A quarter of an hour ago, it was…

The time is…

d)

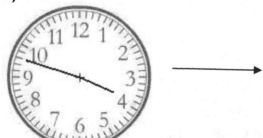

The time is…

Draw a clock to show:
In one hours time, it will be…

e)

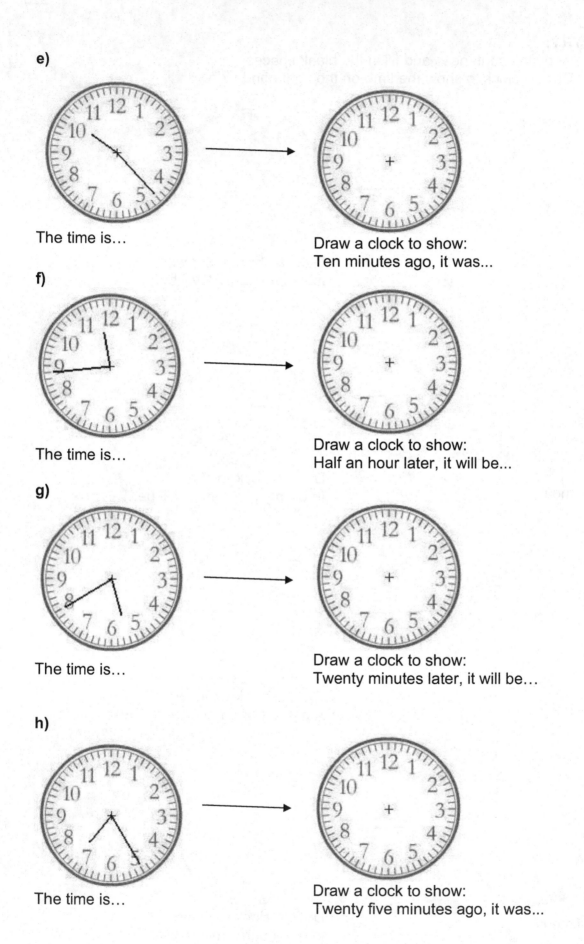

The time is…

Draw a clock to show:
Ten minutes ago, it was...

f)

The time is…

Draw a clock to show:
Half an hour later, it will be...

g)

The time is…

Draw a clock to show:
Twenty minutes later, it will be…

h)

The time is…

Draw a clock to show:
Twenty five minutes ago, it was...

Digital Time

Digital format shows the time using **numbers**.

Here are some examples:

Example 1: Write this analogue time into digital time.

Answer:

- The minute hand points to the **5**.

    Remembering the minutes go up in multiples of **5**:

    5 x 5 = 25 minutes.

- The short hand points in between **one** and **two**.

    It is **past one** and has not reached **two**.

    The digital time is **1:25**.

a.m. and p.m.

- **a.m.** tells us the times from **midnight to noon**.
  In this time, you are asleep.
  This is also before you have lunch!

- **p.m.** tells us the times from **noon to midnight**.
  In this time, you eat lunch and dinner.
  This is also when you go to sleep!

We write these after the digital time (and in words for analogue) to tell us what **time of the day** it is.

<u>**Example 2:**</u>   The time is **twenty two to eleven** in the **morning**.

Write this time in digital format.

<u>**Answer:**</u>

- At the **7th** hour, there are:

  5 x 7 = 35 minutes out of the whole hour

On the clock, the minute hand points **three lines** after the 7.

We know **one line** represents **one minute**.

So **three lines** represents **three minutes**.

Three lines after the 7 is:

  35 + 3 = 38 minutes

- The short hand points in between **ten** and **eleven**.

  It is **past ten** and has not reached **eleven**.

  The digital time is **10:38**.

- In the question, it tells us that the time is in the **morning**.

  Add **a.m.** to our time:
  10:38 a.m.

YOU TRY:

3. Write these times shown on the clocks into **digital** format.
    Don't forget to write <u>a.m.</u> and <u>p.m.</u>

a)

Afternoon

b)

Before noon

c)

Night

d)

Morning

e)

Sunset

f)

Early morning

g)

Evening

h)

Dawn

4. Use a compass to draw some clocks into your books.
Draw the **hands** on your clocks to show the correct time.

a) Quarter to four

b) Quarter to two

c) Eleven o'clock

d) Half past three

e) Six o'clock

f) Quarter past twelve

g) Half past eight

h) Quarter to ten

5. How would these times appear in numbers (digital)?

a) Quarter to seven

b) Half past five

c) Quarter past one

d) Nine o'clock

e) Half past four

f) Ten to two

g) Four past five

h) Twenty two to six

i) Fourteen past eleven

j) Nineteen to ten

k) Eleven past twelve

l) Twenty five to three

6. Write these times in words:

a) 5:15

b) 7:45

c) 9:30

d) 2:00

e) 11:17

f) 8:48

g) 6:25

h) 4:52

i) 3:31

j) 1:19

k) 12:01

l) 10:53

Section 3
Topic 3: Time Facts

OBJECTIVES

In this module, you will:

❖ Be able to convert between <u>seconds</u> and <u>minutes</u>

❖ Be able to convert between <u>minutes</u> and <u>hours</u>

KEYWORDS:

Seconds   Unit of time (60 seconds = 1 minute)

Minutes   Unit of time (60 minutes = 1 hour)

Hours   Unit of time

PREVIOUS KNOWLEDGE:

❖ Units – Length

❖ Units – Mass and Volume

Time conversions

We know <u>seconds (secs)</u>, <u>minutes (mins)</u>, and <u>hours (h)</u> are all <u>units of time</u>.

Likewise to previous work with units, we can convert between these different units of time.

Converting between seconds and minutes

$$\div 60$$

60 seconds = 1 minute

$$\times 60$$

<u>Example 1</u>:   How many seconds are there in 5 minutes?

<u>Answer</u>:   We are converting from <u>mins</u> to <u>secs</u>.

So we must **multiply by 60**:

5 mins x 60 = <u>300 **secs**</u>

<u>Example 2</u>:   How many minutes are there in 240 seconds?

<u>Answer</u>:   We are converting from <u>secs</u> to <u>mins</u>.

So we must **divide by 60**.

240 secs ÷ 60 = <u>4 **mins**</u>

YOU TRY:

1. Follow the example in order to change these minutes into seconds.

a) 2 mins : _____2 mins x 60_____ = _120_ secs

b) 5 mins   h) 2.5 mins

c) 7 mins   i) 4.5 mins

d) 6 mins   j) 12 mins

e) 9 mins   k) 15 mins

f) 10 mins   l) 20 mins

g) 1.5 mins

2. Follow the example in order to convert these times in seconds into minutes.

a) 120 secs : _____120 secs ÷ 60___ = __2__ mins

b) 180 secs   h) 150 secs

c) 240 secs   i) 210 secs

d) 60 secs   j) 270 secs

e) 300 secs   k) 30 secs

f) 420 secs   l) 90 secs

g) 360 secs

3. Follow the example to help you write these seconds into both <u>minutes and seconds</u>:

a) 190 seconds = __180__ + __10__ = _____3 mins 10 secs_____

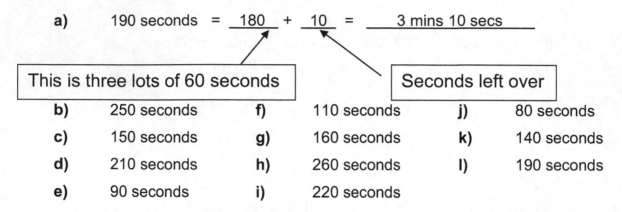

This is three lots of 60 seconds | Seconds left over

b) 250 seconds   **f)** 110 seconds   **j)** 80 seconds

c) 150 seconds   **g)** 160 seconds   **k)** 140 seconds

d) 210 seconds   **h)** 260 seconds   **l)** 190 seconds

e) 90 seconds   **i)** 220 seconds

4. It takes the lady at the cafeteria 280 seconds to make 4 sandwiches. How long does she take in <u>minutes and seconds</u> to make <u>one sandwich?</u>

5. It takes Larry 100 seconds to correct each of his wrong answers.
He has 5 incorrect answers.
How long does he take in <u>minutes and seconds</u> to correct them all?

6. It takes 9.5 minutes to listen to the playlist on my iPod.
How many <u>minutes and seconds</u> is this?

7. On average, it takes 40 seconds to catch a fish.
Barry caught five fish that day.
In <u>minutes and seconds</u>, how long was he fishing for?

Converting between minutes and hours

$$\div 60$$

60 minutes = 1 hour

x60

Example 3:   How many minutes are there in 2 hours?

Answer: We are converting from <u>hours</u> to <u>mins</u>.

So we must **multiply by 60**.

2 hours x 60 = <u>120 **mins**</u>

Example 4:   How many hours are there in 180 minutes?

Answer: We are converting from <u>mins</u> to <u>hours</u>.

So we must **divide by 60**.

180 mins ÷ 60 = <u>3 **hours**</u>

Example 5:   Larry has <u>5</u> lessons on Monday.

Each lesson lasts <u>30</u> minutes.

How many <u>hours and minutes</u> does he spend in lessons?

Answer: In lessons, he spends:

<u>5</u> x <u>30</u> = <u>150 minutes</u>

To convert this into hours, we need to **divide by 60**.

150 mins ÷ 60 = <u>2 hours 30 mins</u>

YOU TRY:

8.   Follow the example in order to change these hours into minutes.

a)   2 hours : _____2 hours x 60_____ = _120_ mins

b)   3 hours   **h)**   1.5 hours

c)   8 hours   **i)**   5.5 hours

d)   7 hours   **j)**   11 hours

e)   4 hours   **k)**   20 hours

f)   10 hours   **l)**   15 hours

g)   2.5 hours

9.   Follow the example in order to convert these times in minutes into hours.

a)   120 mins : _____120 secs ÷ 60_____ = __2__ hours

b)   180 mins   **h)**   150 mins

c)   240 mins   **i)**   210 mins

d)   60 mins   **j)**   270 mins

e)   300 mins   **k)**   30 mins

f)   420 mins   **l)**   90 mins

g)   360 mins

10.   Follow the example to help you write these minutes into <u>hours and minutes</u>:

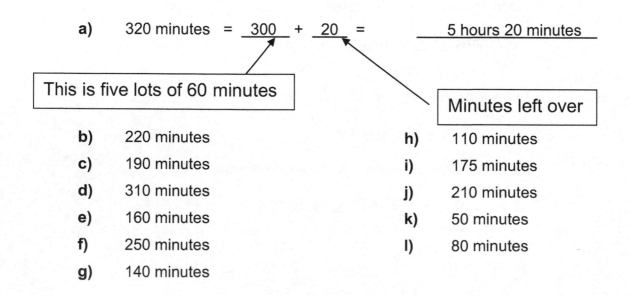

a)   320 minutes = _300_ + _20_ =   _____5 hours 20 minutes_____

This is five lots of 60 minutes

Minutes left over

b)   220 minutes   **h)**   110 minutes

c)   190 minutes   **i)**   175 minutes

d)   310 minutes   **j)**   210 minutes

e)   160 minutes   **k)**   50 minutes

f)   250 minutes   **l)**   80 minutes

g)   140 minutes

11. Alan has 7 lessons.

    Each lesson lasts 30 minutes.

    How long does Alan spend in lessons in <u>hours and minutes</u>?

12. I spend 25 minutes running each day.

    How many <u>hours and minutes</u> do I run in total each week?

13. An overnight train journey lasts 7.5 hours.

    How long is this in <u>hours and minutes</u>?

14. Abi works 6 hours each day in a beauty salon.

    Every client has a forty minute treatment.

    How many clients could Abi see in a day?

Section 3
Topic 4: Timetables and Calendars

OBJECTIVES

In this module, you will:

❖ Cover <u>hours</u>, <u>days</u>, <u>months</u>, <u>weeks</u> and <u>years</u>

❖ Learn how to solve problems on the above

❖ Be able to read <u>calendars</u> and find <u>dates</u> on them

❖ Understand that the <u>columns</u> of a <u>timetable</u> represent <u>different</u> trains and their departure and arrival times.

❖ Be able to <u>read</u> and <u>use</u> timetables

KEYWORDS:

Hours	There are 24 hours in a day
Days	There are 7 days in a week and 365 days in a year (to be more accurate, there are 365 ¼ days in a year)
Weeks	There are 52 weeks in a year
Leap year	A year where February has 29 days instead of 28
Fortnight	2 weeks
Annual	1 year
Decade	10 years
Century	100 years
Millennium	1000 years
Calendars + Timetables	Tables that help us to plan future events

PREVIOUS KNOWLEDGE:

❖ Time

Hours, Days, Weeks, and Years

> ➤ There are <u>24 hours</u> in <u>1 day</u>.

> 7 There are <u>7 days</u> in <u>1 week</u>.

> 8 There are <u>365 days</u> in <u>1 normal year</u>.

> ➤ There are <u>52 weeks</u> in <u>1 year</u>.

Here are some important terms you should know:

Word	Meaning
Fortnight	2 weeks
Annual	1 year
Decade	10 years
Century	100 years
Millennium	1000 years

<u>Example 1:</u>   How many <u>hours</u> are in <u>two days</u>?

<u>Answer:</u>   There are <u>24 hours</u> in <u>one day</u>.

In <u>two days</u> there are:

2 x 24 = <u>48 hours</u>

<u>Example 2:</u>   How many <u>days</u> are there in <u>three weeks</u>?

<u>Answer:</u>   There are <u>7 days</u> in <u>one week</u>.

In <u>three weeks</u> there are:

7 x 3 = <u>21 days</u>

YOU TRY:

Copy and fill in the spaces for the following:

1.

a)   One day is _____ hours.

b)   Two days is _____ hours.

c)   Four days is _____ hours.

d)   Half a day is _____ hours.

e)   One week is _____ hours.

2.

a)   In two weeks, there are _____ days.

b)   In a fortnight, there are _____ days.

c)   In five weeks, there are _____ days.

d)   In seven weeks, there are _____ days.

e)   In ten weeks, there are _____ days.

f)   In nine weeks, there are _____ days.

g)   In fifteen weeks, there are _____ days.

3.

a)   A <u>century</u> is _____ years, so <u>half a century</u> is _____ years.

b)   A <u>decade</u> is _____ years. <u>Three decades</u> is _____ years.

c)   A <u>millennium</u> is _____ years.

Months and Leap Years

    a) There are <u>12 months</u> in <u>one year</u>.

    b) <u>Months</u> either have <u>30 days</u> or <u>31 days</u>.

    c) <u>February</u> is special and usually has <u>28 days</u>.

    d) In a <u>leap year</u>, February has <u>29 days</u>.

    e) In general, if the year is <u>divisible by 4</u>, then it is a leap year.

<u>Example 1</u>:   I save £200 a month.
                            How much do I save in a year?

<u>Answer</u>:   There are <u>12 months</u> in <u>one year</u>.

                          So in <u>one year</u>:

                          £200 x 12 = <u>£2400</u>

<u>Example 2</u>:   Kenny earns £6972 annually.
                            How much does he earn in <u>1 month</u>?

<u>Answer</u>:   "Annually" means "every year".

                          There are <u>12 months</u> in <u>one year</u>.

                          Therefore, in <u>one month</u>:

                          £6972 ÷ 12 = <u>£581</u>

<u>Example 3</u>:   Will the year 2020 be a leap year?

<u>Answer</u>:   2020 will be a leap year if it is divisible by 4 (that is, there is **NO** remainder)

                          2020 ÷ 4 = <u>505</u>

                          It is divisible by 4, so 2012 will be a leap year.

4. Copy the table and find out the number of days in each month.

Month	Number of Days
January	31
February (usually)	
March	
April	
May	
June	
July	
August	
September	
October	
November	
December	

5. Copy and answer the following questions.

a) Find out how many days are there in a <u>normal</u> year.

b) Find out how many days are there in a <u>leap</u> year.

c) A <u>normal</u> year has _____ whole weeks and _____ days left over.

d) A <u>leap</u> year has _____ whole weeks and _____ days left over.

e) Will the year 2035 be a leap year?

6.

a) I save £100 a month.

How much do I save in a year?

b) Alan's <u>annual</u> salary is £8532.

How much does he earn every month?

c) Jessica earns £250 a week.

How much does she earn in a year?

Calendars and Dates

Calendars tell us the <u>days</u> of the month so we can <u>plan</u> activities.
Here's the calendar for June 2011:

June

Mon	Tue	Wed	Thu	Fri	Sat	Sun
		1	2	3	4	5
6	7	8	9	10	11	12
13	14	15	(16)	17	18	19
20	21	22	23	24	25	26
27	28	29	30			

We write the circled date as: <u>Thursday 16.06.11 or 16/06/11</u>

<u>Note</u>:
Here in Britain, we put the day number before the number of the month. In the USA (or looking at a lot of info on the internet) they would write the data above as 6/16/11.

Let's look at some examples of questions for this month of the calendar.

<u>Example 1</u>:   How many Tuesdays are there in June 2011?

<u>Answer</u>:   We are looking at the second column:

June

Mon	Tue	Wed	Thu	Fri	Sat	Sun
		1	2	3	4	5
6	7	8	9	10	11	12
13	14	15	16	17	18	19
20	21	22	23	24	25	26
27	28	29	30			

Just counting, there are <u>four</u> Tuesdays, the 7th, 14th, 21st and 28th in June 2001.

Example 2:   What day is it on the 26th?

Answer: Find 26 on the calendar:

June

Mon	Tue	Wed	Thu	Fri	Sat	Sun
		1	2	3	4	5
6	7	8	9	10	11	12
13	14	15	16	17	18	19
20	21	22	23	24	25	26
27	28	29	30			

26th is in the column for **Sunday**.

Example 3:   What day is two days after the 25th?

Answer: Look at the 25th:

June

Mon	Tue	Wed	Thu	Fri	Sat	Sun
		1	2	3	4	5
6	7	8	9	10	11	12
13	14	15	16	17	18	19
20	21	22	23	24	25	26
27	28	29	30			

One day after, it is the 26th, which is a Sunday.

In two days time, we reach the 27th, which is a Monday.

YOU TRY:

7.   Use this Calendar to answer the following questions:

July 2021

Mon	Tue	Wed	Thu	Fri	Sat	Sun
			1	2	3	4
5	6	7	8	9	10	11
12	13	14	15	16	17	18
19	20	21	22	23	24	25
26	27	28	29	30	31	

a)   How many Tuesdays are there?

b)   How many Sundays are there?

c)   What is the 2<u>nd day</u> of the month?

d)   What is the <u>last day</u> of the month?

e)   What day is it on the 28th?

f)   What day is it on the 11th?

g)   What day is it on the 3rd?

h)   What day is 27.07.21?

i)   What day is 12.07.21?

j)   My friends and I are going to the cinema on the 14th.
      What is the full date?

k)   Tina has an appointment for her hair on the first Sunday of July.
      Write the full date of this Sunday of the month.

8.   Using the full **2021** Calendar, write down the <u>days</u>:

a)   7th of July   (Answer: <u>Wednesday</u>)

b)   21st of June

c)   2nd of February

d)   9th of May

e)   11th of December

f)   30th of August

g)   5th of January

h)   22nd of April

i)   One day after 22nd April

j)   One day before 7th January

k)   Three days after 12th October

l)   Four days before 16th November

m)   Three days before 11th December

n)   One week after 12th October

o)   One week before 18th of May

p)   A fortnight after 1st of January

9.   Using the **2021** Calendar, write down the following <u>dates</u>:

The <u>second</u> Thursday of August. (Answer: 12.08.21)

a) The <u>first</u> Tuesday of December

b) The <u>third</u> Saturday of January

c) The <u>fourth</u> Monday of May

d) The <u>last</u> Monday of April

e) The <u>first</u> Wednesday of September

f) The <u>third</u> Sunday of February

g) Halloween

h) Christmas

i) Boxing Day

j) Valentine's Day

k) St. Patrick's Day

l) Remembrance Day

Reading Timetables

Let's have a look at this timetable:

STATION	Departure Times	
	Train 1	Train 2
King's Cross	09:06	09:23
Potters Bar	09:21	09:39
Hatfield	09:27	-
Stevenage	09:42	09:47
Hitchin	09:47	09:52

This column tells us about the first train.

This column tells us about the second train.

- Each time in the timetable represents what time the train stops, arrives and leaves the station.

- A box without a time ("-") means the train does not stop at that station.

Example 1:   The second train departs King's Cross and eventually reaches Hitchin.

What stations does it stop at?

Answer: We are looking at the second column of departure times.

It stops in Potters Bar at 09:39.

It does not stop in Hatfield.

It stops in Stevenage at 09:47.

The stations it stops by on the way to Hitchin are Potters Bar and Stevenage.

Example 2:   What time does the first train arrive in Potters Bar?

Answer: The <u>first column</u> of <u>departure</u> times is the <u>first train</u>.

Find Potters Bar Station on the left side:

STATION	Departure Times	
	Train 1	Train 2
King's Cross	09:06	09:23
Potters Bar	09:21	09:39
Hatfield	09:27	-
Stevenage	09:42	09:47
Hitchin	09:47	09:52

The first train arrives in Potters Bar at <u>09:21</u>.

Example 3:   Christopher is in King's Cross and needs to be at Hatfield by ten o'clock.

Which train should he catch?

Answer: The first train arrives in Hatfield at 09:27.

The second train does not stop at Hatfield.

STATION	Departure Times	
	Train 1	Train 2
King's Cross	09:06	09:23
Potters Bar	09:21	09:39
Hatfield	09:27	-
Stevenage	09:42	09:47
Hitchin	09:47	09:52

Christopher must catch the <u>first train</u> in order to get to Hatfield before ten o'clock.

YOU TRY:

10.   Use the timetable for two trains to find the answers to the following
       questions.

STATION	Departure Times	
	Train 1	Train 2
Luton	10:04	10:25
Leagrave	-	10:28
Harlington	-	10:34
Flitwick	10:14	10:38
Bedford	10:28	10:50

a)   What time does the <u>first train</u> stop in Flitwick?

b)   What time does the <u>second train </u>stop in Flitwick?

c)   What time does the second train <u>depart</u> Luton?

d)   What time does the 10:04 train <u>arrive</u> in Bedford?

e)   What time does the second train <u>stop</u> at Leagrave?

f)   The 10:04 train <u>does not</u> stop at two stations. What stations are they?

g)   The first train departs Luton and eventually reaches Bedford.
      What is the only station it stops at on this journey?

h)   Jason needs to get to Harlington before 11:00.
      Will he be able to catch the first train?

i)   Jessica and her friends arranged to meet up in Bedford at half past ten in the
      morning.
      What train should she take to ensure she isn't late?

j)   I walked to Flitwick and arrived there at 10:30.
      What time is the next train?

k)   What station do both trains eventually arrive at?

11. Use the timetable for three trains to find the answers to the following questions.

STATION	Departure Times		
	Train A	Train B	Train C
St Helier	-	08:50	09:17
Mitcham Eastfields	08:28	08:58	09:24
Mitcham Junction	08:31	09:01	-
Hackbridge	08:34	09:04	09:30
Carshalton	08:37	-	09:33
Sutton	08:40	09:10	09:36

a) What time does the first train <u>arrive</u> in Carshalton?

b) The second train <u>departs</u> St Helier at what time?

c) What time does the 08:28 train <u>arrive</u> in Sutton?

d) The first train leaves Hackbridge at what time?

The next train to arrive at the same station is at what time?

e) Which train <u>does not</u> stop at Carshalton?

f) The third train <u>does not</u> stop at which station?

g) Unlike the other two trains, the first train starts its journey from which station?

h) Name the three stations that <u>all</u> three trains stop at.

Copy and complete the following:

i) The first station the second train stops at is _____.

j) The second station the 09:17 train stops at is _____.

k) The 08:50 train is currently at Mitcham Eastfields. It stops at _____ and _____ before reaching Sutton.

l) The third train is currently at Hackbridge.

How many more stations does it stop at before reaching Sutton?

m) I arrived in St Helier at 9:00.

What time is the next train?

Section 3
Topic 5: Adding and Subtracting Time Problems

OBJECTIVES

In this module, you will:

❖ Learn what we mean by **duration**

❖ Be able to find durations of times

❖ Be able to <u>add and subtract</u> durations

❖ Be able to <u>simplify</u> answers when we have more than 60 minutes in our answer

KEYWORDS:

Duration   "How long" sometimes takes.

e.g. the duration of a bus journey is how long it takes when you get on and when you get off the bus

PREVIOUS KNOWLEDGE:

❖   Adding and Subtracting
❖   Time
❖   Time Facts

Duration

The word 'duration' means how long something takes.

In order to find the duration, we must remember there are 60 minutes in 1 hour, and also remember our work on **Time Facts**.

There are two examples to follow in this module:

Example 1:   Find the duration from 8:30am to 9:50am.

Step 1:   **Find the hour duration**

The start minutes are smaller than the end minutes.

The number of hours is:

9 – 8 = 1 hours

Step 2:   **Find the minute duration**

The start time shows 30 minutes.

To make up 50 minutes in the end time, we need:

50 – 30 = 20 minutes

Step 3:   **Write the duration**

The duration is 1 hour and 20 minutes.

<u>Example 2:</u>	Find the duration from 7:40am to 10:20am.

Step 1: **<u>Find the hour duration</u>**

The <u>start minutes</u> are <u>greater</u> than the <u>end minutes</u>.

This time, we must <u>subtract an extra 1 hour</u>:

$10 - 7 - 1 = $ <u>2 hours</u>

Step 2a: **<u>Find the minute duration</u>**

We know there are 60 minutes in an hour.

The start time shows 40 minutes.

To <u>make up 60 minutes</u>, we need:

$60 - 40 = $ <u>20 minutes</u>

Step 2b: The <u>20 minutes</u> we found just now takes us <u>up to the hour</u>.

As the end time shows 20 minutes, we need an <u>extra</u> 20 minutes:

$20 + 20 = $ <u>40 minutes</u>

Step 3: **<u>Write the duration</u>**

The duration is <u>2 hours and 40 minutes</u>.

Example 3:   Gary does a training session on two days.
               He likes to know how long he trains for.

               Work out how many minutes he trains on those days.

Day	Start	Finish
Tuesday	8:00am	9:10am
Friday	8:40am	9:30am

Answer: Let's look at each day individually:

Tuesday   Starts from 8:00am and finishes at 9:10am.

               This follows the method for Example 1.

Step 1:   The end minutes are greater than the start minutes.
               9 – 8 = 1 hours

Step 2:   Subtract the end minute with the start minutes:
               10 – 00 = 10 minutes

Step 3:   Duration: 1 hours 10 minutes

Friday   Starts from 8:40am and finishes at 9:30am.

               This follows the method for Example 2.

Step 1:   The end minutes are smaller than the start minutes.
               9 – 8 – 1 = 0 hours

Step 2a: Find how many minutes make up to the hour.
               60 – 40 = **20** minutes

Step 2b: Find how many more minutes we need:
               20 + 30 = 50 minutes

Step 3:   Duration: 0 hours 50 minutes

YOU TRY:

1. Follow the method for Example 1 to find the duration of these questions.
    a)   10:00am to 11:30am

    b)   2:30pm to 3:40pm

    c)   4:20pm to 5:30pm

    d)   9:10am to 10:40am

    e)   1:15am to 2:45am

    f)   Twenty past six to ten to nine

    g)   Quarter past nine to quarter to one

    h)   Twenty five to eight to quarter to nine

2. Follow the method for Example 2 to find the duration of these questions.

    a)   9:50am to 10:10am

    b)   3:40pm to 4:00pm

    c)   5:20pm to 8:10pm

    d)   9:50am to 12:30pm

    e)   7:55am to 9:05am

    f)   Twenty to six to twenty five past eight

    g)   Five past two to four o'clock

    h)   Half past three to ten past six

3. Niall is a doctor.
Here is his list of appointments.

How much time does he spend with each patient?
Copy and complete the table.

Appointment Time	Patient	Length of Appointment
10:00am	Miss Haris	
10:10am	Alex Edwards	
10:30am	Mr Tucker	
10:45am	Holly Spencer	
11:00am	Mrs Wesley	
11:10am	Hayley Becks	
11.50am	Jared Strong	
12.20pm	Miss Louise	
1.00pm	Lunch Break	
1.30pm	Fiona Parker	
1.45pm	Finish	

Adding and Subtracting Time

We can add and subtract time values normally, but sometimes we must simplify our answers using our knowledge about time.

Reminder: 60 seconds = 1 minute

                   60 minutes = 1 hour

Example 1:   1hr 20min + 2hr 30min.

Answer: Add the hours and minutes separately:

                        Hours: 1 + 2 = 3
                        Minutes: 20 + 30 = 50

                    The total time is 3hr 50min.

Example 2:   3hr 15min + 1hr 55min.

Answer: Add the hours and minutes separately:

                        Hours: 3 + 1 = **4**
                        Minutes: 15 + 55 = **70**

                    We have **70** minutes, but there are 60 minutes in 1 hour.

                    We can change **70 minutes** into **1hr 10min**.

                    Now we need to add **1** more hour to what we worked out before:

                        New hours: 4 + 1 = 5

                    The total time is 5hr 10min.

Example 3:   2hr 30min – 1hr 10min

Answer: The start minutes are greater than the end minutes.

                    We can just subtract normally the hours and minutes separately:

                        Hours: 2 – 1 = 1
                        Minutes: 30 – 10 = 20

                    The difference in time is 1hr 20min.

Example 4:   2hr 20min – 1hr 30min

Answer: The start minutes are smaller than the end minutes.

                  This time, convert 1 hour to 60 minutes from the start time:

                      2hr 20min becomes **1hr 80min**.

                  They are the same but are written differently.

                  We are borrowing an hour from the start time, like how we borrow
                  from the tens or hundreds when we're subtracting numbers.

We now have: **1hr 80min** – 1hr 30min

                  Subtract the hours and minutes separately again:

                      Hours: 1 – 1 = 0
                      Minutes   80 – 30 = 50

                  The difference in time is 0hr 50min.

YOU TRY:

4.   Use the method in Example 1 to find these totals:

a)   1hr 20min + 2hr 30min   **e)**   5hr 5min + 1hr 25min

b)   2hr 10min + 1hr 40min   **f)**   2hr 50min + 1hr 5min

c)   4hr 20min + 2hr 10min   **g)**   6hr 5min + 8hr 20min

d)   3hr 15min + 4hr 15min

h)   Three of Darius's favourite shows last 1hr 10min, 2hr 5min and 1hr 30min. What is the total time?

i)   Fiona says she works 2hr 10min in the morning, 1hr 30min in the afternoon and 3hr 5min in the evening. How long did she work altogether that day?

5.   Use the method in Example 2 to simplify your answers:

a)   2hr 30min + 1hr 40min   **e)**   3hr 15min + 1hr 50min

b)   1hr 50min + 1hr 20min   **f)**   5hr 30min + 6hr 35min

c)   4hr 20min + 2hr 40min   **g)**   7hr 35min + 3hr 45min

d)   2hr 25min + 1hr 45min

h)   Niall took part in the triathlon. He took 1hr 20min to swim, 2hr 50min to cycle and 20min to run. How long did Niall take to complete the triathlon?

i)   The journey time from Upminster to King's Cross is 40min. The journey time from King's Cross to Kennington is 25min. What is the total journey from Upminster to Kennington?

6.   Use the method in Example 3 to find the differences in time:

a)   1hr 50min - 1hr 10min   **e)**   7hr 25min - 1hr 10min

b)   2hr 40min - 1hr 20min   **f)**   8hr 30min - 3hr 25min

c)   5hr 40min - 4hr 15min   **g)**   3hr 25min - 1hr 5min

d)   6hr 15min - 2hr 15min

h)   At home, Paul has 5 hours 30 minutes of free time.

He spends 3 hours 15 minutes doing his homework.

How much free time does he have left to play?

i)   Tina works 3 hours 20 mins in the morning up to lunch.

So far, she's worked 2 hours 10 minutes.

How much longer does she have until lunch?

7.   Use the method in Example 4 to find the differences:

a)   2hr 20min - 1hr 30min   **f)**   7hr 30min - 6hr 45min

b)   7hr 10min - 4hr 50min   **g)**   8hr 20min - 5hr 40min

c)   3hr 10min - 1hr 30min

d)   4hr 15min - 2hr 20min

e)   6hr 25min - 1hr 35min

h)   Paul's mother watches a comedy show followed by drama.

She spends 1hr 20min watching altogether.

If the comedy show lasts 45min, how long is the drama?

i)   Thomas spends in total 4hr 30 mins doing both his homework and playing. He spends 1hr 40min playing.

How long does he spend on his homework?

Pre 11+ Maths

Summer Half Term Test

Full Name: ………………… Test Date: ……………….…............

Tutor: …………………………. Day of Lessons: ……………….………

This test lasts 30 minutes.

Read the questions carefully and try all the questions.
If you cannot do a question, move on to the next one.
If you have finished, check over your answers.

The total marks for this paper is 45.

1. Solve the following, showing all your working.

a) $\frac{7}{26} + \frac{5}{13}$ (2)

b) Total 20195, 5106 and 39. (2)

c) Find the product of thirty four point six, and seven. (2)

d) i) Find the HCF of 8 and 38. (1)

ii) Hence simplify the fraction $\frac{8}{38}$. (1)

2. Paul weighs 36.25 kg and Harry weighs 40.80 kg.

What is the difference in their weights? (2)

3. Write in the missing numbers in the spaces for the scale. (3)

a)

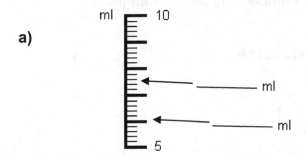

b)

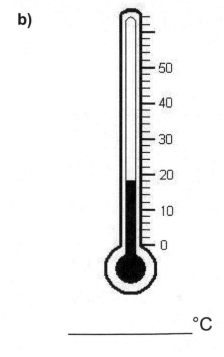

_____°C

4. Find the perimeter of the following shape. (2)

_____

_____

Answer: _____ cm

5. Larry's been given homework about <u>mass</u> and <u>volume</u>.

Unfortunately, some of the words and numbers are faded from the photocopying.

Help him work out what the numbers should be. (2)

a)

1000 ml = 1 ℓ

x 1000

b)

÷1000

g = 1 kg

x1000

c) Hence find the following:

i) 9.42 ℓ = _____ ml (1)

ii) 482g = _____ kg (1)

6.   Find:

a)   70% of 230.   (2)

b) $\frac{9}{11}$ of 77.   (2)

7.   How would these times appear in 12 hour digital format?
(don't forget to include am and pm).

a)   Twenty three to five in the afternoon.   _____   (1)

b)   Seventeen past one in the morning.   _____   (1)

8.   Sarah has 8 lessons at school.
Each lesson lasts 35 minutes.
How many minutes does she spend in lessons?   (1)

_____

Write your answer in hours and minutes   (1)

_____

9. Use the timetable to help to answer the following questions.

STATION	Departure Times		
Littleport	08:02	08:25	08:39
Waterbeach	08:11	08:34	-
Cambridge	08:20	08:43	08:59
Liverpool Street	-	-	09:45
King's Cross	-	09:43	-

a) Which train does not stop at Waterbeach? (1)

b) The 08:39 train arrives at Cambridge at what time? (1)

c) For the 08:25 train, how long is the journey between Littleport and King's Cross? (1)

d) I arrive at Waterbeach at 08:13.
How many minutes do I need to wait for the next train? (1)

10. My sister and I want to both watch TV.

My sister's program finishes at 08:46pm, but I want to watch my film at 8:23pm.

a) How much of my film will I miss? (1)

b) My film finishes in 2 hours and 34 minutes.

Write the time it ends. (2)

11. By car, the journey to the theme park is 3 hours and 23 minutes.

By train, the journey is only 1 hour and 45 minutes.

a) How much shorter in time is the train compared to the car? (2)

_____

_____

_____

b) Last Saturday, there was a lot of traffic.

It took an extra 54 minutes by car.

What was the total journey time? (1)

_____

_____

c) On that Saturday, we left at 7:21am.

What time did we arrive at the theme park by car that day? (2)

_____

_____

d) What time would we have reached the theme park if we took the train on the same Saturday?

_____

_____ (1)

12.   Here is a shape made up of two combined rectangles.

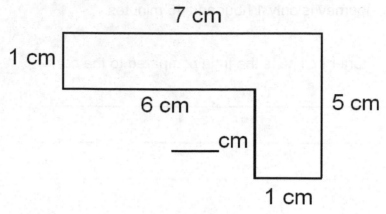

a)   Find the missing length and write it in.   (1)

_____

b)   Calculate the perimeter of this shape.   (1)

_____

_____

_____

Answer _____ cm

c)   Find the area of this shape.   (3)

_____

_____

_____

_____

Answer _____ cm²

Section 3
Topic 7: 12 and 24 Hour Clocks

OBJECTIVES

In this module, you will:

❖   Be able to use **<u>a.m.</u>** and **<u>p.m.</u>** to describe the time of day.

❖   Understand the difference between the 12 hour and 24 hour format of time.

❖   Be able to convert from 12 hour to 24 hour format.

❖   Be able to convert from 24 hour to 12 hour format.

❖   Be able to use 12 and 24 hour formats in problem solving.

KEYWORDS:

a.m.	Times that are from midnight to noon.
p.m.	Times that are from noon to midnight.
12 hour format	A digital time using 1 to 12 for hours.
24 hour format	A digital time using 0 to 24 for hours. It uses <u>4 digits</u>.

PREVIOUS KNOWLEDGE:

❖   Time
❖   Time Facts
❖   Add/ Subtract Time Problems

| **Prenote**: | Both 12 hour and 24 hour clock formats are **digital times**. |

12 hour clock

This format of time uses only the hours 1 to 12.

We must use **a.m.** and **p.m.** to specify what <u>half</u> of the day we are talking about.

- **a.m.** (**a**nte **m**eridian) tells us the times from <u>midnight to noon</u>.

  Words to describe this time: <u>early morning</u>, <u>dawn</u>, <u>morning</u>, <u>before afternoon</u>

- **p.m.** (**p**ost **m**eridian) tells us the times from <u>noon to midnight</u>.

  Words to describe this time: <u>afternoon</u>, <u>sunset</u>, <u>evening</u>, <u>night</u>

<u>Example 1</u>:   1:34 a.m. tells us it is just past midnight.

<u>Example 2</u>:   7:25 p.m. tells us it is evening.

24 hour clock

This format of time uses the hours 0 to 24, and **does not** use a.m. or p.m..

To be consistent, we always write 24 hour clock times using <u>four digits</u>.

- The times <u>from the 0th hour up to the 12th hour</u> are the times from <u>midnight to noon</u>.

- The times <u>from the 12th hour up to the 24th hour</u> are the times from <u>noon to midnight</u>.

<u>Example 3</u>:   07:52 tells us it is early morning.

<u>Example 4</u>:   00:24 tells us it is night.

YOU TRY:

1. Copy and complete the sentences by filling in <u>a.m.</u> or <u>p.m.</u> in the spaces so that the times make sense.

a) Breakfast served from 7:00___ to 9:30___.

b) The library opens at 8:30___ to 5:30___.

c) We're closed for lunch from 12:30___ to 1:15___.

d) Jessica has a Dentist appointment at 2:45___.

e) Sunrise is at 4:35___.

f) Sunset is at 8:20___.

g) My exam starts at 10:00___ and finishes at 11:00___.

h) Harry is having a party between 1:30___ and 3:00___.

i) Abi goes to sleep at 11:00___.

j) On Sunday, Sarah goes shopping at 12:20___.

---

2. Copy and link these times of the day with <u>a.m.</u> or <u>p.m.</u>.

Afternoon

Midnight

a.m.

Morning

p.m.

Dawn

Prenote:	When converting between 12 and 24 hour formats, we only look at the hours.
	The minutes remain <u>unchanged</u>.

Converting from 12 hour to 24 hour clock

You must recognise whether the time is <u>a.m.</u> or <u>p.m.</u>

Also make sure the resulting 24 hour clock time has <u>four digits</u>.

- If it is <u>a.m.</u>, <u>write a 0 before the hour</u> when necessary.

- If it is <u>p.m.</u>, we must <u>add 12 hours</u> to the time.

<u>Example 1</u>:   Change 5:30 a.m. into the 24 hour clock.

<u>Answer</u>:   Here it is a.m., so we follow the first bullet point.

        To make up <u>four digits</u>, we must add a 0 before the hour:

        5:30a.m. is <u>05:30</u> in the 24 hour clock.

<u>Example 2</u>:   Change 7:25 p.m. into the 24 hour format.

<u>Answer</u>:   This time it is p.m., so we follow the second bullet point.

        We add 12 hours:

        $7 + 12 = \underline{19}$

        Therefore:

        7:25p.m. is <u>19:25</u> in a 24 hour format.

273

YOU TRY:

3.   Change these 12 hour times into 24 hour format.

a)	2:52 a.m.		**n)**	3:33 a.m.
b)	6:30 p.m.		**o)**	4:50 p.m.
c)	8:25 p.m.		**p)**	6:32 p.m.
d)	7:12 a.m.		**q)**	9:49 a.m.
e)	1:56 p.m.		**r)**	2:32 p.m.
f)	4:23 a.m.		**s)**	7:57 p.m.
g)	9:50 a.m.		**t)**	8:20 a.m.
h)	12:40 a.m.		**u)**	1:11 p.m.
i)	9:02 p.m.		**v)**	9:42 a.m.
j)	5:30 p.m.		**w)**	7:55 a.m.
k)	6:12 a.m.		**x)**	5:39 p.m.
l)	8:35 p.m.		**y)**	4:30 a.m.
m)	7:32 a.m.		**z)**	7:36 a.m.

Converting from 24 hour to 12 hour clock

Here, we must recognise whether the 24 hour time refers to the <u>first half</u> of the day or the <u>second half</u>.

- If the hours are from <u>0 up to 12</u>, it is the <u>first half</u> of the day.

  This time of day is <u>a.m.</u>. We <u>remove the 0 in front</u>.

- If the hours are from <u>12 up to 24</u>, it is the <u>second half</u> of the day.

  This time of day is <u>p.m.</u>. Here, we <u>subtract 12</u> from the hours.

Example 1:   Write 03:25 on the 12 hour clock.

Answer: The hours are between 0 and 12.

        This tells us it is the <u>first half</u> of the day, so we use <u>a.m.</u>.

        Remember to remove the front 0:

        03:25 is <u>3:25 a.m.</u> on the 12 hour clock.

Example 2:   Write 16:20 in 12 hour format.

Answer: The hours are between 12 and 24.

        This tells us it is the <u>second half</u> of the day, so we use <u>p.m.</u>.

        Subtract 12 from the hours:

        $16 - 12 = \underline{4}$

        16:20 is <u>4:20 p.m.</u> in the 12 hour format.

YOU TRY:

4. Change these 24 hour times into 12 hour format.

Remember to write a.m. and p.m. to specify the time of day.

a)	01:23		**n)**	06:12
b)	05:54		**o)**	17:27
c)	13:10		**p)**	18:24
d)	15:40		**q)**	08:08
e)	10:30		**r)**	09:32
f)	06:28		**s)**	19:05
g)	15:30		**t)**	17:23
h)	20:52		**u)**	16:45
i)	19:19		**v)**	23:55
j)	09:45		**w)**	00:41
k)	06:49		**x)**	14:22
l)	23:20		**y)**	01:49
m)	00:10		**z)**	07:30

5. <u>Copy</u> and find the new times.

2 hours later

a) 00:50 $\longrightarrow$ _____

45 minutes later

b) 12:35 $\longrightarrow$ _____

30 minutes earlier

c) 15:40 $\longrightarrow$ _____

3 hours before

d) 12:30 $\longrightarrow$ _____

50 minutes after

e) 05:40 $\longrightarrow$ _____

15 minutes earlier

f) 09:10 $\longrightarrow$ _____

2 hours earlier

g) 3:45p.m. $\longrightarrow$ _____ (a.m. or p.m.?)

1 hour later

h) 2:30a.m. $\longrightarrow$ _____ (a.m. or p.m.?)

20 minutes later

i) 9:50p.m. $\longrightarrow$ _____ (a.m. or p.m.?)

40 minutes before

j) 7:35p.m. $\longrightarrow$ _____ (a.m. or p.m.?)

in 25 minutes

k) 8:10a.m. $\longrightarrow$ _____ (a.m. or p.m.?)

5 hours ago

l) 1:00a.m. $\longrightarrow$ _____ (a.m. or p.m.?)

6. Paul walked to King's Cross Station and arrived there at 4:20p.m

a) What is this time on the 24 hour clock?

The timetable tells him the next train comes in 35 minutes.

b) What time will it arrive in the 12 hour format?

c) What is this time in the 24 hour format?

7. Fiona's exam starts at 11:00a.m. and lasts for 2 hours.

a) Write in 24 hour format the time she finishes.

b) Write the same time in the 12 hour format.

8. The lunch queue is estimated to take 15 minutes.

a) If I queue up at 1:20p.m., what time will I get lunch?

Write your answer in the 12 hour format.

b) Today, I finished my lesson early and started queuing at 1:05pm.

Write the time I get lunch this time in the 24 hour format.

9. In both the 12 and 24 hour format, it is confusing to describe exactly midnight and midday as 12:00 or 00:00. To remove this confusion, we add or remove 1 minute to these times. Copy and complete:

12 hour clock:

a) The two times to describe midday are _____ or _____.

b) The two times to describe midnight are _____ or _____.

24 hour clock:

c) The two times to describe midnight are _____ or _____.

(Note that there isn't any confusion for the midday time)

Section 3
Topic 8: 2D Shapes

OBJECTIVES

In this module, you will:

❖ Understand about 2-D shapes and polygons.

❖ Know the properties of and be able to recognise the three special triangles.

❖ Know the properties of and be able to recognise special quadrilaterals.

❖ Understand regular and irregular shapes, and about each of their interior angles.

❖ Understand the notation used for angles, right angles, equal sides.

KEYWORDS:	
2-D	Two dimensional – shapes you can draw on your paper.
2-D Polygon	Two dimensional shapes with straight sides
Parallel	Two line are parallel when they are always the same distance apart and will never meet.
Right angle	An angle of 90°.
Interior angle	Any specified angle inside a shape.
Regular shape	A shape with equal interior angles and equal sides.
Irregular shapes	A shape that does not have equal interior angles or equal sides.

2-D Shapes and Polygons

2-D means "two dimensional". These are shapes you can draw flat onto your paper.

The name of a 2-D is determined by the number of sides.

2-D Polygons are 2-D shapes with straight sides.

Special Triangles

Triangles have three sides. There are three special triangles.

Note:
If two or more lines have one short line going through them, this tells us that the lines have the same length

The triangles below have curved lines in them. The curved lines tell us that those angles are the same.

Name	Shape	Angle properties	Side Properties
Equilateral Triangle		Equal angles	Equal sides
Isosceles Triangle		Two equal angles	Two equal sides
Scalene Triangle		No equal angles	No equal sides

YOU TRY:

1. <u>Copy</u> these sentences and use the words to fill in the blank spaces.

    Equal   Equilateral   Sides   Isosceles   Angles

a)   A triangle with equal sides is called an _____ triangle.

b)   A triangle with two equal angles is called an _____ triangle.

c)   A scalene triangle does not have any equal _____ or _____.

d)   An equilateral triangle has _____ sides.

2.   Write the names of these triangles.

a)

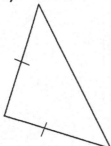

b)

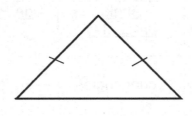

c)

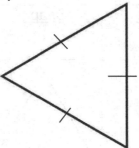

d)

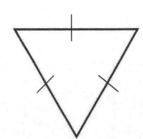

e)

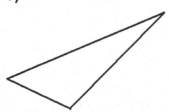

f)

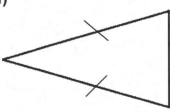

Special Quadrilaterals

Quadrilaterals have <u>four sides</u>. There are a few special quadrilaterals.

<u>Note</u>:

<u>Parallel</u>: when two lines are always the same
              distance apart and will never meet.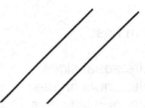

<u>Right angle</u>: an angle of 90º.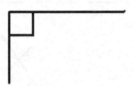

<u>Special Quadrilaterals</u>:

<u>Square</u>

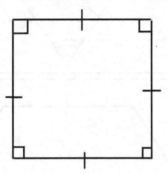

A square has:

> Equal sides.
> Equal angles, each being a right angle.
> Opposite parallel sides.

<u>Rectangle</u>

A rectangle has:

> Opposite equal sides.
> Equal angles, each being a right angle.
> Opposite parallel sides.

Parallelogram

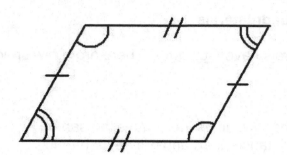

A parallelogram has:

➢   Opposite equal sides.
➢   Opposite equal angles.
➢   Opposite parallel sides.

Rhombus

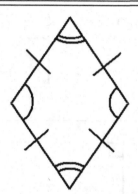

A rhombus has:

➢   Equal sides.
➢   Opposite equal angles.
➢   No parallel sides.

Trapezium

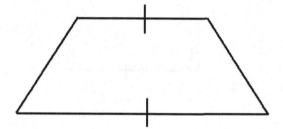

For a shape to be a trapezium, it must have <u>only</u> one pair of parallel sides.

Kite
A kite has:

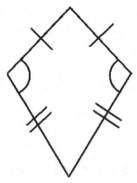

➢   Two pairs of equal sides
➢   One pair of equal angles
➢   No parallel sides.

YOU TRY:

3.   <u>Copy</u> these sentences and use the words to fill in the blank spaces.

Angles   Equal   Opposite   No   Right
Parallel   Sides   Two   One

a)   A square has _____ sides and _____. Each angle is a _____ angle. The _____ sides are also parallel.

b)   A rectangle also has equal _____ and _____ opposite parallel _____. However, unlike a square, it has _____ pairs of equal sides.

c)   The parallelogram has equal opposite _____ and _____. The opposite _____ are also _____.

d)   Like a square, a rhombus has _____ sides. It has opposite equal _____ and two pairs of _____ sides.

e)   Trapeziums are the only shapes that do not focus on its _____. Instead, it only has one pair of parallel _____.

f)   A kite is the only shape that does not have any relation with opposite _____. It has _____ pairs of equal sides, _____ pair of equal angles and _____ parallel sides.

4. Write the names of these quadrilaterals.

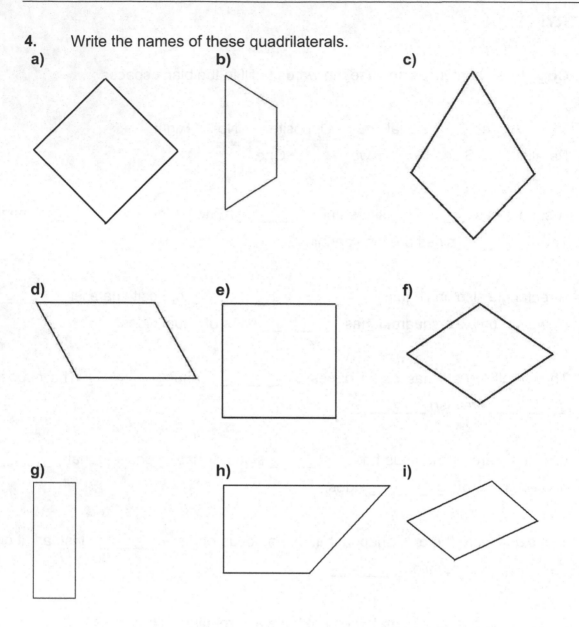

Regular and Irregular Shapes

Regular shapes have:

➢   Equal interior angles (all angles inside the shape are the same)
➢   Equal side lengths.

Here are a few:

Name	Shape	Number of sides	Interior angles	Each interior angle
Regular Triangle		3	180°	180° ÷ 3 = 60°
Regular Quadrilateral		4	360°	360° ÷ 4 = 90°
Regular Pentagon		5	540°	540° ÷ 5 = 108°
Regular Hexagon		6	720°	720° ÷ 6 = 120°

Irregular shapes have different interior angles OR different side lengths. Here are some examples:

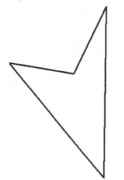

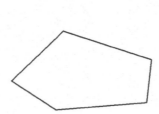

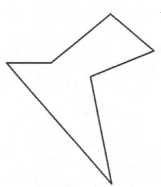

Irregular quadrilateral   Irregular pentagon   Irregular hexagon

5. <u>Copy</u> and fill in the blank spaces.

a) A regular shape has equal _____ and equal _____.

b) A triangle has _____ sides. In an _____ triangle, each angle is 60°. In every triangle, the total interior angles add up to 180°.

c) A _____ has 4 sides. Each angle is ____° in a square, and in any quadrilateral, the total _____ angles add up to 360°.

d) In every pentagon, the total interior angles add up to ____°, and have 5 _____. For a regular pentagon, each interior _____ is 108°.

e) The name we give a shape with 6 sides is a _____. Every angle is 120° in a regular _____, and the total of the angles is ____°.

6. Name these shapes.
Choose from the words [regular or irregular] together with [triangle, quadrilateral, pentagon, or hexagon].

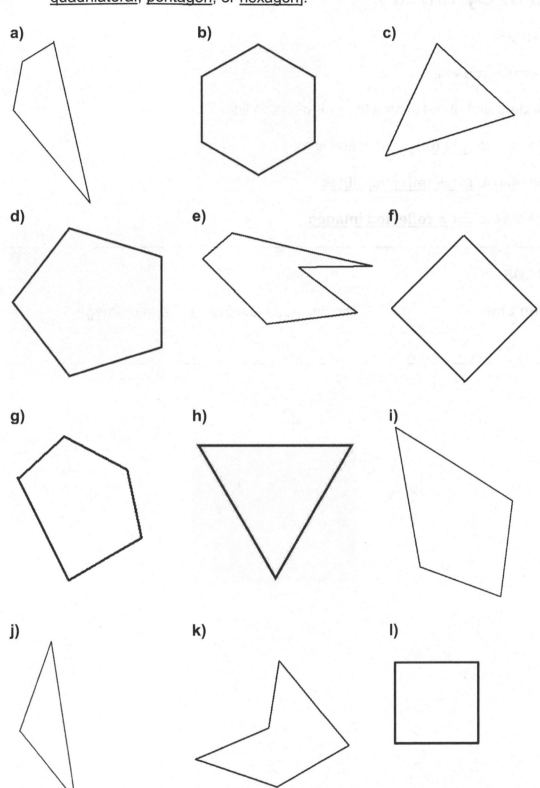

a)

b)

c)

d)

e)

f)

g)

h)

i)

j)

k)

l)

Section 3
Topic 9: Symmetry

OBJECTIVES

In this module, you will:

❖ Understand about symmetry and reflection lines

❖ Be able to **recognise** reflection lines.

❖ Be able to **draw reflection lines**

❖ Be able to **draw reflected images**.

KEYWORDS:

Reflection Line   The line used to mirror a shape or image.

Symmetry

There are three main types of symmetry: <u>translation</u>, <u>rotation</u> and <u>reflection</u>.

In this module, we will be looking at <u>reflection</u>.

Recognising Reflection Lines

In order to reflect something, we need the <u>line of reflection</u>.

To see reflection lines, you must check whether the <u>two sides</u> are the same. You can think of them as mirror lines.

In other terms, it <u>cuts the shape in **exactly** in half.</u>

Example 1:   Which of these lines are lines of symmetry?

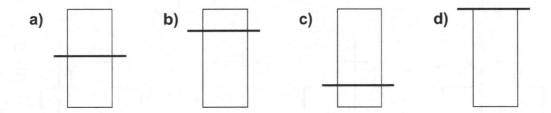

a)   b)   c)   d)

<u>Answer:</u>

We must look at which one of these cuts the shape exactly in half.

The line of symmetry in **a)** cuts the rectangle in half.
It is the same <u>top and bottom.</u>

YOU TRY:

1. Which of these lines are lines of symmetry?

   Write down the correct letter.

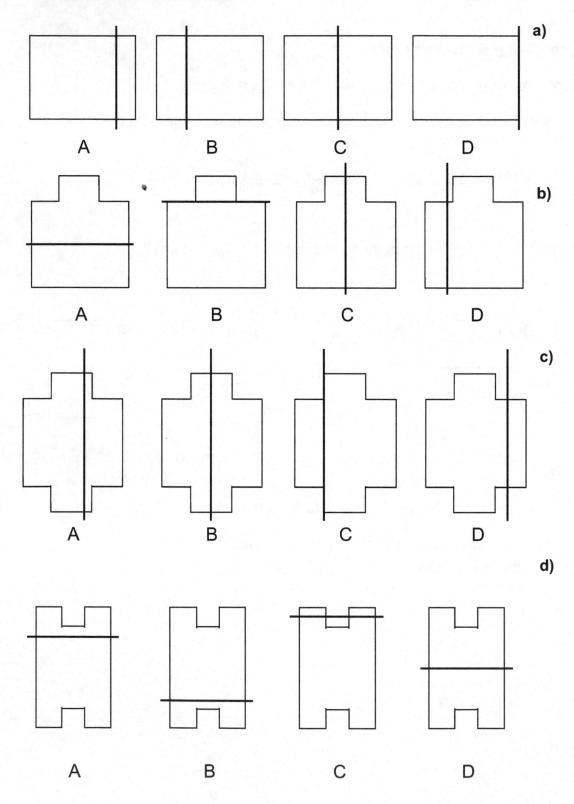

a)

A   B   C   D

b)

A   B   C   D

c)

A   B   C   D

d)

A   B   C   D

e)

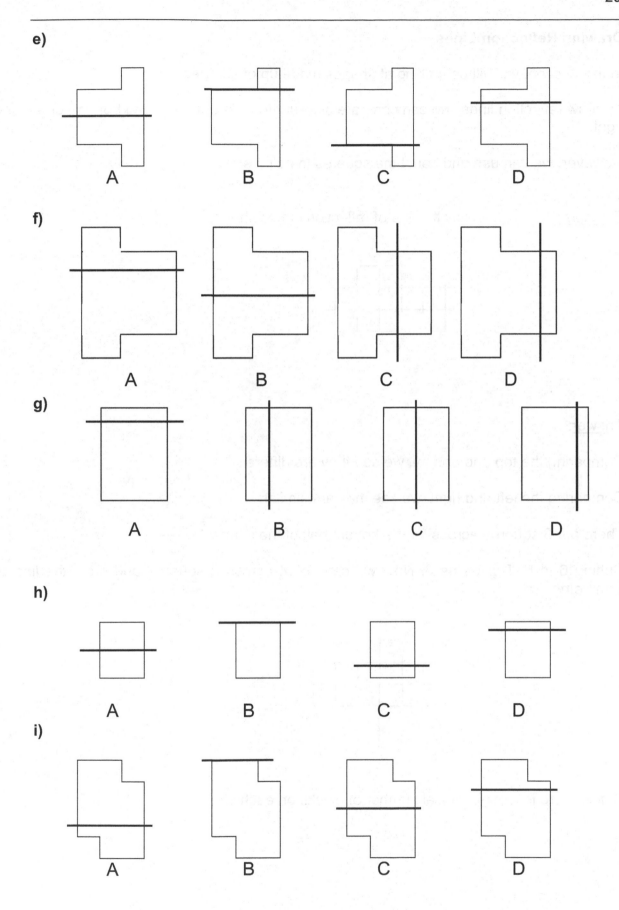

A   B   C   D

f)

A   B   C   D

g)

A   B   C   D

h)

A   B   C   D

i)

A   B   C   D

Drawing Reflection Lines

In this section, we will be looking at shapes made up of squares.

To draw reflection lines, we can compare similarly to before the top and bottom, or left and right.

However, we can use and count the squares to help us.

Example 1:   Draw the line of reflection of this shape:

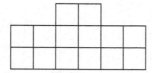

Answer:

Comparing the top and bottom, we see they are different.

Comparing the left and right, we see they are similar.

There are 6 squares across for the bottom half of the shape.

Cutting 6 in half gives us 3. Now we can count across 3 squares and draw the line of symmetry:

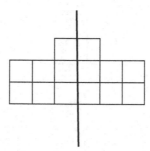

Notice there is a symmetrical number of blocks on each side.

<u>Example 2:</u>   Draw the line of reflection of this shape:

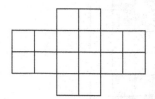

<u>Answer</u>:

Comparing the top and bottom, we see they are similar.

There are 6 squares across for the middle half of the shape.

Cutting 6 in half gives us 3. Now we can count across 3 squares and draw the line of symmetry:

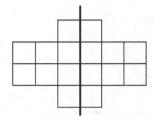

Similarly to the last example, there is a line of symmetry across as well.

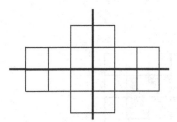

Notice again there is a symmetrical number of blocks on each side.

This shape has <u>2</u> lines of symmetry.

Example 3:   Draw the line of reflection of this shape:

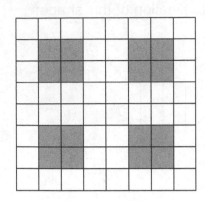

Answer:

Comparing the top and bottom, we see they are similar.

Each shape is made up of 4 squares with 2 squares in between each one.

Cutting 2 in half gives us 1. Now we can count down 1 square from the top shape and draw the line of symmetry:

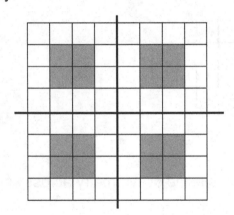

Similarly for the left and right:

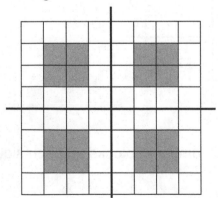

Notice again there is a symmetrical number of blocks on each side.

YOU TRY:

2. <u>**Copy**</u> these shapes and draw in the lines of symmetry:

a)

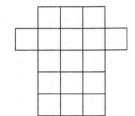

e)

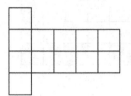

i)

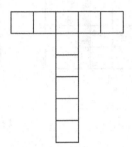

b)

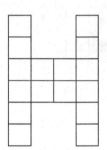

f)

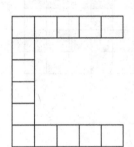

j)

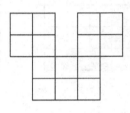

c)

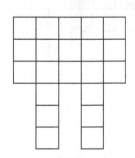

g)

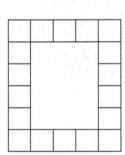

k)

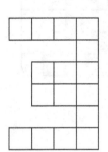

d)

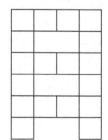

h)

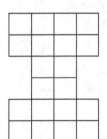

l)

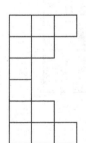

m)

q)

u)

n)

r)

v)

o)

s)

w)

p)

t)

x)

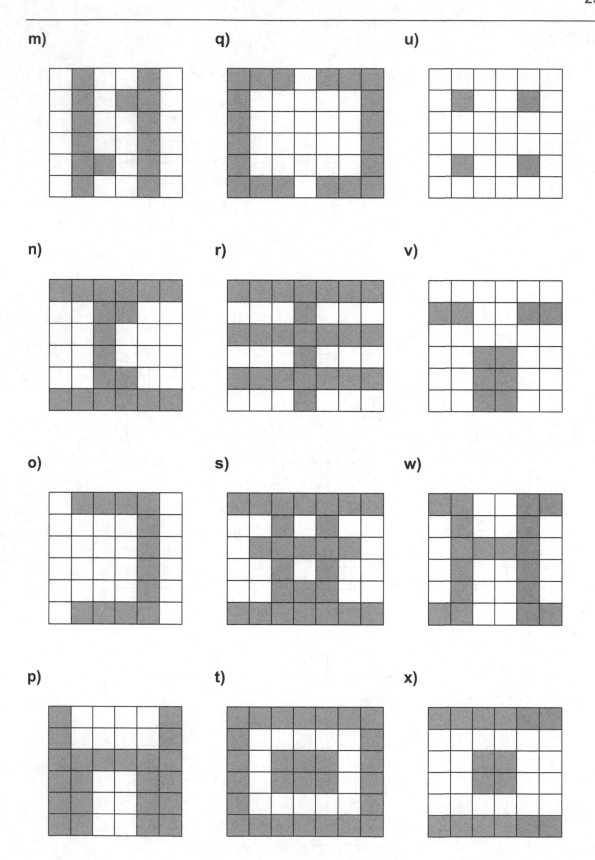

Drawing Reflection Images

<u>Example 1:</u>   Draw the reflected image:

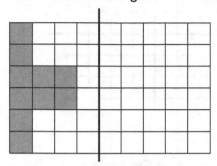

<u>Answer:</u>

Count how many squares it takes each box of the shape to get to the line:

Now count this number of squares to the right of the line:

Now draw the reflected shape:

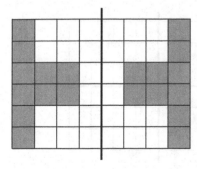

Example 2:   Draw the reflected image:

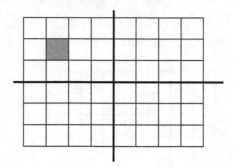

Answer:

Following exactly the same method as Example 1:

Count how many squares it takes the shaded box to get to the line:

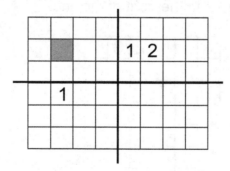

Now count this number of squares to the right of the line:

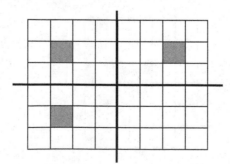

Now draw the reflected shaded boxes:

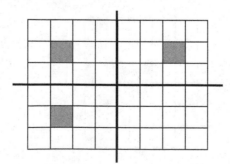

Repeat again to find the reflections of the shapes we found:

Counting the squares up to the line:

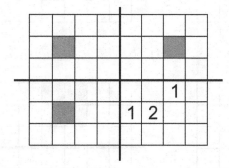

Counting the squares past the line:

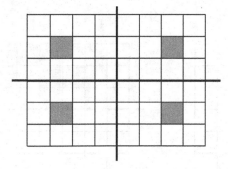

The reflected shape in these two lines of symmetry looks like this:

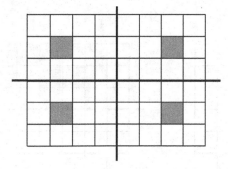

YOU TRY:

3.   <u>Copy</u> these shapes and find the reflected shape.

       <u>Shade</u> in your reflected shape.

a)

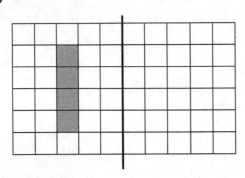

b)

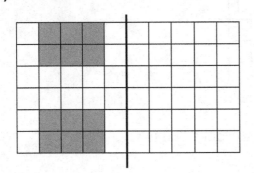

c)

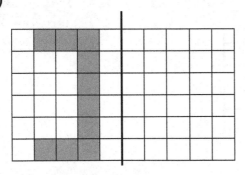

d)

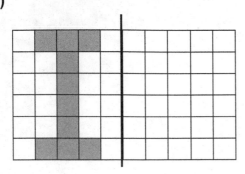

e)

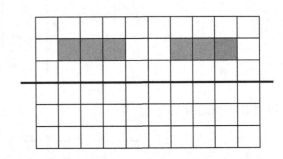

f)

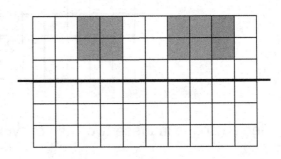

g)

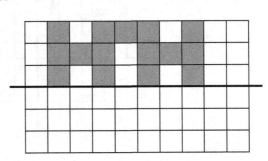

h)

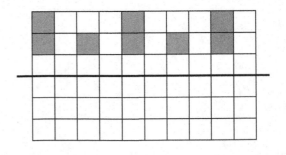

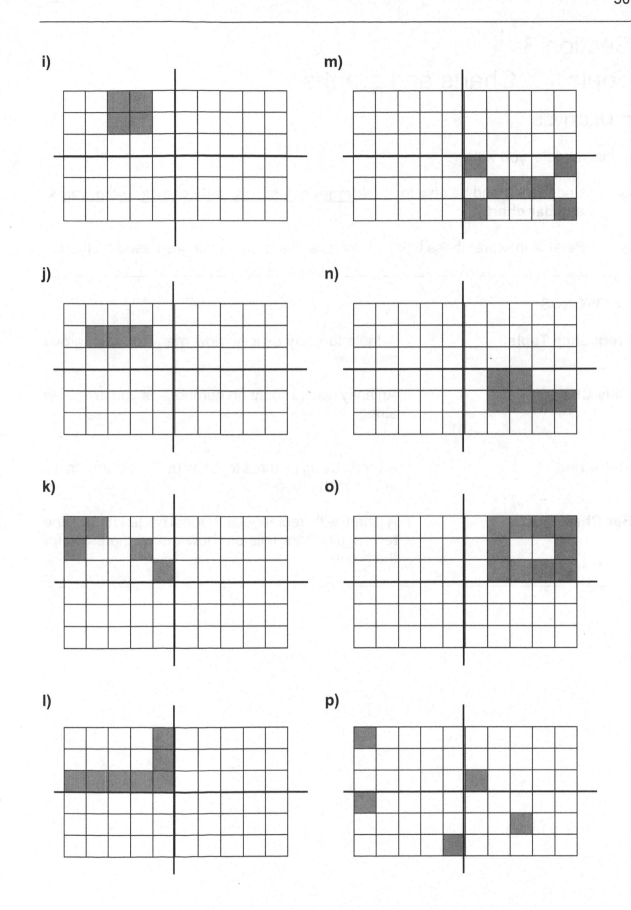

Section 3
Topic 10: Charts and graphs

OBJECTIVES

In this module, you will:

❖   Understand and be able to draw **frequency tables**, **tally charts**, **pictograms** and **bar charts**

❖   Be able interpret these types of charts and answer questions based on them

KEYWORDS:

Frequency Table	A table to show exactly 'how many' of each group there are.
Tally Chart	An easy way to count up numbers as you are going along.
Pictogram	A graph using pictures to show us 'how many' there are of each group.
Bar Chart	A chart with rectangular bars. The lengths of the rectangular bars tells us 'how many' there are for the group.

Frequency Tables

The word 'frequency' tells us 'how many'.

Frequency tables are tables that tell us exactly how many of each group there are.

Example 1:   A class of 15 did a survey on their favourite colours.

Blue   Green Blue   Yellow Red

Red   Yellow Green Yellow Blue

Green Red   Blue   Red   Red

Fill in the following frequency table for the colours.

Colour	Number
Blue	
Red	
Yellow	
Green	

Answer: Count the number each colour appears and write this into the table.

Blue appears 4 times.
Red appears 5 times.
Yellow appears 3 times.
Green appears 3 times.

Writing these into the table:

Colour	Number
Blue	4
Red	5
Yellow	3
Green	3

Tally Charts

In order to help us count the total number for <u>large</u> sets of data, we use <u>tallies</u>.

We draw tallies like this:

1 |

2 | |

3 | | |

4 | | | |

5   ~~| | | |~~

6   ~~| | | |~~ |

12   ~~| | | |~~ ~~| | | |~~ | | etc.

For every 5, we draw a <u>diagonal line</u> and continue as normal.

<u>Example</u>:   Draw up a tally chart to show the number of times each of the following numbers appears.

    2   5   4   2   4   3   2   2   5   4

    3   4   1   2   3   4   1   2   3   2

<u>Answer</u>:   Counting up the numbers:

     1 appears <u>2</u> times. The tally is | | .

     2 appears <u>7</u> times. The tally is ~~| | | |~~ | | .

     3 appears <u>4</u> times. The tally is | | | | .

     4 appears <u>5</u> times. The tally is ~~| | | |~~ .

     5 appears <u>2</u> times. The tally is | | .

It will be useful to include the frequency in the chart.

Draw up the tally chart:

Number	Tally	Frequency
1	||	2
2	|||| ||	7
3	||||	4
4	||||	5
5	||	2

YOU TRY:

1. Copy the table in each question and fill in the tally and frequency columns for the different sets of data in each question.

a) Here are the colours sold at GSP Motors last week:

White	Black	Red	Red	Blue	Black
Blue	White	Blue	White	Black	Red
Red	Red	Black	Black	White	Blue

Colour	Tally	Frequency
White		
Black		
Red		
Blue		

b) George goes bird watching every Monday morning.
    Here is the number of birds he spotted over the past weeks:

3	6	5	4	6	5	3	4	3	6	5
4	4	5	6	4	3	3	6	4	3	5

Number	Tally	Frequency
3		
4		
5		
6		

For the following, create your own table to include your tallies and frequencies.

c) Temperatures of London afternoons:

12° 14° 10° 10° 14° 12° 13° 14° 10° 10°

14° 13° 11° 12° 12° 13° 14° 11° 10° 13°

d) The number of test questions Larry completed every day:

20 23 19 22 23 19 20 21 22 22

23 19 19 20 21 23 22 21 20 20

e) The children's favourite colours in Class 5C:

Purple Green Red   Purple Pink Red

Green Purple Green Pink   Pink   Green

Red   Green Red   Purple Red   Red

f) The number of goals scored every match by Harry in the tournament:

0   1   4   3   1   2   0   0   2   0   3

2   1   1   2   0   2   1   1   0   0   2

g) The number of songs I listen to every day:

36   39   37   40   39   36   36   38   39   40   38

38   37   38   36   38   37   39   40   40   37   36

h) The number of movies Jessica rented each month:

18   16   18   16   19   17   18   18   19   17   18

19   18   17   18   17   19   19   17   16   17   18

Pictograms

Pictograms use <u>pictures</u> to tell us <u>how many</u> there are.

A <u>key</u> tells us <u>how many</u> each picture represents.

<u>Example</u>:

In a recent tournament, Flint scored 4 goals, Gary scored 2 goals and Wayne scored 1 goal.

Complete the pictogram to show how many goals were scored by the players:

Name	Goals scored
Jack	
Flint	
Paul	⚽ ⚽ ⚽
Gary	
Wayne	

<u>Key</u>: Each ⚽ symbol stands for <u>one goal</u>.

<u>Answer</u>:　　The key tells us each football represents <u>one goal</u>.

Therefore:

Flint:　4 goals is shown as:　 ⚽ ⚽

Gary: 2 goals is shown as:　

Wayne: 1 goal is shown as:

Filling in the table:

Name	Goals scored
Jack	
Flint	
Paul	
Gary	
Wayne	

Example 2:   Find a suitable key for the following data:

20   15   25   20

Answer: Remembering your work on HCF, the highest number that goes into each one is 5.

YOU TRY:

2.   <u>Copy</u> and complete the following pictograms given the key in each question:

a)   The number of DVDs rented yesterday at Joanna's store:

<u>Type of DVD</u>	<u>DVDs rented</u>	<u>Frequency</u>
Comedy	⊙ ⊙	8
Drama	⊙ ◖	6
Horror		4
Action		8
Children's		10
Educational		6
Sports		12
Soap		4

<u>Key</u>:   Each stands for 4 DVDs.

<u>Note</u>:   This means that stands for 2 DVDs.

311

b) The number of ants spotted each day last week outside Jack's house for his school project.

Day	Number of ants	Frequency
Monday	🐜 🐜 🐜 🐜	40
Tuesday	🐜 🐜	20
Wednesday		30
Thursday		20
Friday		40
Saturday		10
Sunday		20

(Each 🐜 stands for 10 ants.)

For the following, create your own pictogram chart to show the information.

Use a suitable key for each one.

<u>Hint</u>:   Use the Highest common factor (HCF) to decide on a key.

c)   The number of cars spotted outside the school gate:

Monday:	20
Tuesday:	30
Wednesday:	25
Thursday:	35
Friday:	30

d)   The number of sweets the children bought:

Anna:	10
Jack	15
Jessica:	10
Terry:	20

e)   The number of votes for House Captain: (<u>hint</u>: draw badges)

Harry:	16
Jared:	24
Linna:	32
Fiona:	28

f)   The number of boxes required each day at work:

Monday:	300
Tuesday:	250
Wednesday:	150
Thursday:	200
Friday:	300

Bar Charts

A <u>bar chart</u> is a chart with <u>rectangular bars</u>.

The <u>length</u> of the rectangular bar tells us '<u>how many</u>'.

A common bar chart has 2 axis:

➤ The one going up is for the <u>frequency</u>.

➤ The one going across is for the <u>class</u> (e.g. colours, numbers etc).

<u>Example</u>: Draw a bar chart for the following frequency table.

<u>Colour</u>	<u>Frequency</u>
Green	6
Blue	3
Red	4
Yellow	2

<u>Answer</u>:

Start by plotting the axis. To decide a suitable scale, have a look at the data. It makes sense to write the <u>frequency</u> in <u>intervals of 1</u>. This will be on our side axis.

Make sure you space out the bottom axis appropriately, and also **label** your axis:

Frequency

6														
5														
4														
3														
2														
1														
0	Green			Blue			Red			Yellow		

Colour

Drawing in the rectangular bars:

Frequency

6														
5														
4														
3														
2														
1														
0	Green			Blue			Red			Yellow		

3. Draw bar charts for the following frequency tables.

For each one, find a suitable scale for the frequency.

a)

Favourite Sport	Frequency
Football	12
Basketball	6
Cricket	6
Tennis	10
Baseball	8

b)

Favourite subject	Frequency
English	5
Maths	10
Science	15
Geography	5
Art	15

c)

Nationality	Frequency
British	16
German	10
French	8
American	6
Spanish	8

d)

Favourite season	Frequency
Spring	80
Summer	120
Autumn	80
Winter	60

Interpreting Graphs

We have looked at lots of types of graphs.
We can look at the information and answer questions on them.

Example: Use this bar chart to answer the following questions.

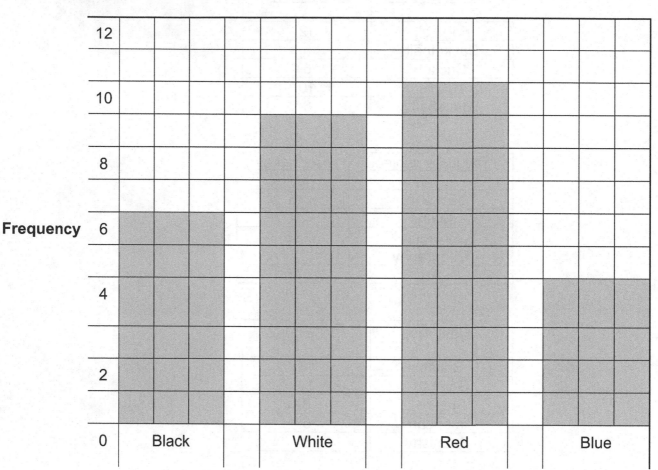

Question: How many blue cars were spotted?

Answer: The 'blue cars' is the last rectangular bar on the far right.

Reading across onto the frequency axis, it lines up with 4.

This means 4 blue cars were spotted.

Question: What was the total number of cars spotted?

Answer: Counting up the total frequency for all the other colours:

There were 6 black cars.
There were 8 white cars.
There were 10 red cars.

In total, there were: 4 + 6 + 8 + 10 = 28 cars

4. Farmer Joe drew a pictogram to show how many strawberries he grew last week.

Day	Strawberries grown	Frequency
Monday	🍓🍓	40
Tuesday	🍓🍓🍓	
Wednesday	🍓🍓🍓	60
Thursday	🍓🍓🍓🍓	
Friday		40

a) Copy the pictogram table.

b) Using the frequency, how much does each strawberry icon represent?

c) Write the frequencies for Tuesday and Thursday for your table.

d) Draw in the missing strawberries for Friday for your table.

e) On which day were most strawberries grown?

f) Were more strawberries grown on Wednesday than Friday?

g) How many more strawberries were grown on Thursday than Wednesday?

h) What was the total number of strawberries grown?

5. The teacher drew a bar chart for a few of his class to compare their test marks:

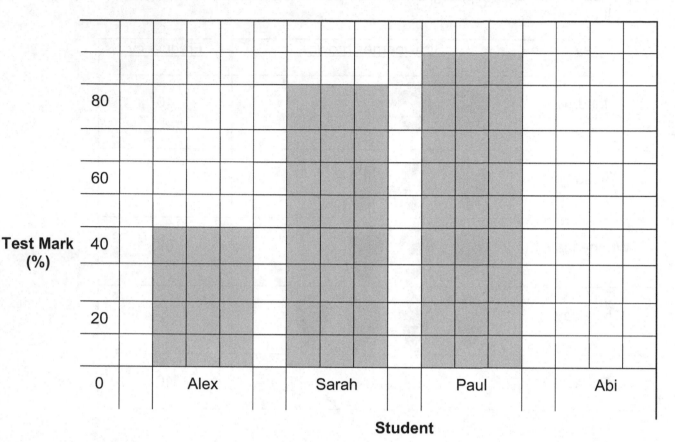

a) Copy the bar chart.

b) Look at the test mark on the left hand side of the bar chart.

Fill in the missing numbers at the top of <u>your</u> bar chart.

c) What mark did Paul get?

d) What mark did Sarah get?

e) How many more marks did Paul get than Alex?

f) Abi achieved 70%. Draw a bar onto <u>your</u> chart to show this.

g) How many students achieved over 50%?

h) Write true or false for the following:

    i) Sarah achieved more than Paul.

    ii) Paul achieved 100%.

    iii) Alex did the worst.

    iv) Abi achieved 30% more than Alex

Section 3
Topic 11: Averages and Range

OBJECTIVES

In this module, you will:

❖ Understand what we mean by the **average** of a set of data.

❖ Be able to find the **average** for a set of data.

❖ Understand what we mean by the **range** of a set of data.

❖ Be able to find the **range** for a set of data.

KEYWORDS:

Average   A measure of the 'middle' value of a set of data.

Range   The difference between the largest
                                         and smallest values in the set of data

PREVIOUS KNOWLEDGE:

❖ Addition and Subtraction
❖ Multiplication and Division

Averages

Averages are a measure of the 'middle' value of a set of data.

For test results, they give us a rough indication on how well the class as a whole is performing.

Let's work through an example to show how we find the average:

<u>Example:</u>   Here are the test results of 5 children:

23   15   25   17   20

Find the average of the results.

<u>Answer:</u>

<u>Step 1:</u>   <u>Add</u> all the test results together:

23 + 15 + 25 + 17 + 20 = **100**

<u>Step 2:</u>   Divide the total by the <u>number of children</u>:

100 ÷ 5 = <u>20</u>

The average mark is <u>20</u>.

YOU TRY:

1. Find the average for these sets of data:

a)	19	20	21		
b)	14	19	12		
c)	12	27	21		
d)	8	2	5	1	
e)	7	2	9	2	
f)	9	8	15	24	
g)	7	4	9	12	8
h)	4	3	1	5	2
i)	8	13	6	4	9
j)	12	2	7	8	3	10
k)	19	1	9	15	14	20
l)	3	2	10	13	9	17

2. Every Monday, Jessica recorded the temperature with her thermometer.

   7°C   13°C   6°C   9°C   10°C

   Find the average of these temperatures.

3. Alan marked the tests of his seven students:

   15   20   27   20   15   24   19

   Find the average test mark.

4. Each week for 5 minutes, Paul counted the number of cars passing outside the school for his project. Here are his results over six weeks:

   12   10   6   9   8   15

   Find the average number of cars that passed by.

5. Fiona gets a different amount of pocket money (in pounds) from her parents each week. Here's what she got the past 4 weeks:

   £18   £17   £25   £20

   Find how much she got on average each week.

Range

The range is the <u>difference</u> between the <u>largest</u> and <u>smallest</u> values in the set of data.

There are two steps to find the range.

<u>Example</u>: Find the range for this set of data:

| 1 | 4 | 9 | 4 | 5 | 10 | 7 | 6 |

<u>Answer</u>:

<u>Step 1</u>: Order the data starting with the smallest:

| 1 | 4 | 4 | 5 | 6 | 7 | 9 | 10 |

<u>Step 2</u>: Find the <u>difference</u> between the <u>largest</u> and <u>smallest</u> values:

$10 - 1 = \underline{9}$

The range of the data is <u>9</u>.

YOU TRY:

6. Find the range for these sets of data:

a)   1   5   0   3   6   8

b)   12   14   19   17   12   20

c)   32   58   23   50   29   33

d)   6   9   3   10   14   7   8

e)   10   18   17   12   24   6   9

f)   24   29   26   22   20   19   31

g)   6   14   9   10   15   12   5   9

h)   4   5   3   9   4   3   7   1

i)   10   18   17   13   24   12   19   20

j)	30	32	38	23	16	34	24	26	26

k)	32	41	42	47	38	31	30	39	38

l)	52	42	59	47	43	51	53	54	58

7. Naveed received his results to his tests today:

   English: 59%
   Maths: 45%
   Science: 79%
   Geography: 50%

   Find the range of his test marks.

8. Last year, Larry decided to measure the length of ants in mm for his school project. Here are some measurements he found:

   2   9   4   3   7   8   7   3   4   2

   Find the range of these ant lengths.

9. The weights of various packs of apples were recorded by the shopkeeper in g.

   250   253   262   254   259   265   271

   Find the range of these weights.

10. Every year, I get some money from my grandparents.
    Over 4 years this is what I got in pounds:

    20.53   29.35   26.39   27.18   32.15

    Find the range of how much I received.

Pre 11+ Maths

Summer End of Term Test

Time allowed: 45 minutes

Full Name: …………………… Test Date: ………………….............

Tutor: ………………………… Day of Lessons: ……………………

Read the questions carefully and try all the questions.
If you cannot do a question, move on to the next one.
If you have finished, check over your answers.

The total marks for this paper is 60.

1) Write the following in figures:

    a) Three thousand, six hundred and five: _____ (1)

    b) Eight thousand and nine:_____ (1)

2) Put the following symbols (=, >, <) between these numbers and (4) fractions.

    a) 6452 ☐ 4532   **b)** 9098 ☐ 9089

    c) $\frac{1}{3}$ ☐ $\frac{1}{2}$ **d)** $\frac{2}{3}$ ☐ $\frac{2}{5}$

3) What is the value of the underlined digit in the following:

    a) 1 <u>4</u> 0 3 7 _____ (1)

    b) <u>6</u> 2 0 4 7 _____ (1)

4) Put these numbers in order of size from the smallest to the largest. (1)

    6520   5601   6205   6052   5260   5620

5) How many 5p pieces are there in £2 ? (1)

                                       Answer _____

6) 3210 ÷ 10 = _____ (1)

7) Calculate $\frac{1}{4}$ of 36 (1)

                                        Answer _____

8) Calculate $\frac{3}{8}$ of 32 (2)

                                        Answer _____

9)   53 + 15 = 90 - ☐ **?**   (1)

Answer _____

10)   Write the missing numbers in the list below.   (2)

(   )   7   11   15   (   )   23

11)   There are 545 children in Mill Hill Academy. 258 of them have school lunch. The rest have a packed lunch.
How many children have a packed lunch?   (1)

Answer _____

12)   Niall saved £519 and Haris saved £397 more than him.
How much has Haris saved?   (1)

Answer _____

13)   Work out the missing digits in the following sums: (6)

a)
```
    3  8  4
 +
    2  6  ☐
  ┌──────────
  ☐  ☐     3
```

b)
```
    2  ☐  0  4
 -
    ☐  1  7  8
  ─────────────
  1  1  ☐     6
```

14)   From the following list of numbers:
54, 39, 64, 89, 13, 20
Write down:
a) The prime numbers _____   (2)

b) The factor of 60 _____   (1)

c) The even numbers _____   (1)

d) The multiples of 3 _____   (1)

15) There are 46 marbles in a bag.   (1)
    a)   How many marbles are in 5 bags?

Answer _____

    b)   How many bags contain 322 marbles?   (1)

Answer _____

16) Apples cost 18 pence each. How much would 8 apples cost?
Write the answer using £'s.

Answer _____ (1)

17)   **a)** $\frac{2}{7} + \frac{3}{5}$

Answer _____ (3)

    b)   1.9 + 4 +12.78

Answer _____ (2)

18) Haris has 56 sweets. He gives $\frac{1}{4}$ of them to his sister. How many sweets does he have left?

Answer _____ (2)

19) It takes Raj 35 minutes to travel to school. What time does he have to leave in the morning to arrive at school at quarter to nine?

Answer _____ (1)

20) Niall's favourite TV show started at 8:25 a.m. and finished at 9:15 a.m. How long does the show last?

Answer _____ (1)

21) This function machine adds two and then multiplies by 3.
Fill in the missing inputs and outputs:

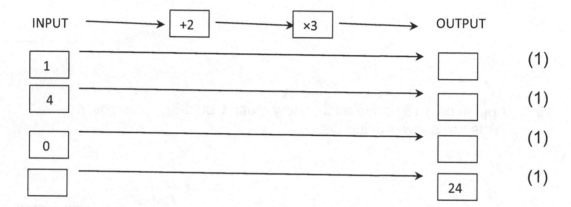

(1)

(1)

(1)

(1)

22) Here are the scales of three different measuring instruments.

a)

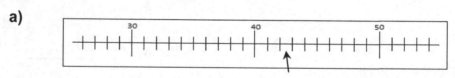

What is the reading indicated by the arrow?

Answer _____ (1)

b)

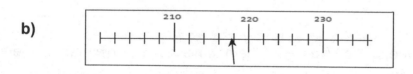

What is the reading indicated by the arrow?

Answer _____ (1)

c)

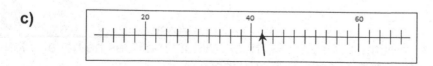

What is the reading indicated by the arrow?

Answer _____ (1)

23) Zirqa went to the cinema. The film starts at 6.35 p.m.
The film lasts 1 hour and 40 minutes. When does it end?

Answer _____ p.m.   (1)

24) Mrs Patel bought 120 balloons for Reena's party.
30% of the balloons were yellow. $\frac{3}{5}$ of the balloons were pink.
The rest were purple.
How many of the balloons were:

a) Yellow   _____   (2)

b) Pink   _____   (2)

c) Purple   _____   (2)

25) Convert the following:

a) How many centimetres (cm) are there in $1\frac{1}{2}$ metres?

Answer _____ (1)

b) How many metres (m) are there in $2\frac{1}{4}$ kilometres?

Answer _____ (2)

c) How many grams (g) are there in 5 kilograms?

Answer _____ (1)

Answers

You can download the detailed answers from our website for £2.99

Printed in the United States
by Baker & Taylor Publisher Services